Life After Midnight

Andrew Parry

Published by Andrew Parry, 2024.

LIFE AFTER MIDNIGHT

First edition. October 4, 2024.

ISBN: 979-8227543783

Written by Andrew Parry.

Table of Contents

Understanding Nuclear Fallout: The Science behind the Danger

Nuclear fallout is the invisible yet omnipresent danger that follows a nuclear explosion. It's not just the initial blast that causes devastation, but the lingering radioactive particles that contaminate the air, water, and soil for months, even years, after the event. To understand how to survive in a world polluted by radiation, it's crucial to comprehend the science behind nuclear fallout and its potential impact on the environment and human health.

When a nuclear explosion occurs, whether it's from a weapon or an accident at a nuclear facility, an immense amount of energy is released in the form of heat and radiation. This blast vaporizes everything in its immediate vicinity and propels a cloud of radioactive debris high into the atmosphere. These particles, composed of highly dangerous isotopes, begin to fall back to Earth in the hours and days following the explosion. This process is what we refer to as nuclear fallout.

The size and spread of fallout depend on various factors: the size of the explosion, the altitude at which the detonation occurred, and local weather patterns, including wind direction and speed. Larger explosions or surface detonations tend to create more fallout, as the blast picks up and disperses debris, dust, and ash from the ground, spreading radiation over a wider area. Winds can carry these particles across hundreds or even thousands of miles, contaminating vast regions far from the original site of the explosion.

Once fallout begins to settle, it clings to everything it touches—buildings, roads, vegetation, and, most dangerously, people. The key threat from nuclear fallout lies in its radioactivity, which comes from isotopes such as Cesium-137, Iodine-131, and Strontium-90. These isotopes emit ionizing radiation, which is capable of damaging or killing living cells by disrupting their DNA. Prolonged exposure to high levels of fallout can cause acute radiation sickness, while even low levels of radiation exposure over time can lead to cancers, genetic damage, and other long-term health effects.

Understanding the concept of radioactive half-life is critical when thinking about fallout. Every radioactive element decays at a specific rate, measured by its half-life—the time it takes for half of the radioactive atoms in a sample to decay into a more stable form. For instance, Iodine-131 has a half-life of about eight days, while Cesium-137 has a much longer half-life of about 30 years. This means that while some isotopes will become less dangerous within a few weeks, others will linger in the environment for decades, continuing to pose a serious risk to life.

The first few hours and days after a nuclear event are the most critical for survival. In the immediate aftermath, the concentration of radioactive particles in the air will be extremely high, and fallout will begin to settle. It's during this time that people need to seek shelter and avoid exposure to the outside environment. A common misconception is that you can immediately leave the fallout area and avoid contamination, but in many cases, the surrounding areas may be equally or even more dangerous due to wind patterns spreading the radioactive particles. The most effective way to survive is to remain indoors in a sealed environment for as long as possible during the initial fallout period.

The radiation from fallout is strongest when it first settles. As time passes, the radiation levels will decrease, but it won't disappear entirely. For example, after 24 hours, the radiation level may be reduced by 80%, but this still leaves significant amounts of radiation in the environment. Understanding this decay process helps survivalists decide when it might be safer to venture outside for supplies or to check the condition of their surroundings. However, without proper testing equipment, it's difficult to gauge whether the radiation levels have dropped to a safe level.

The immediate health effects of exposure to fallout depend on several factors, including the level of radiation, the duration of exposure, and whether the radioactive particles were inhaled, ingested, or absorbed through the skin. High doses of radiation in a short period can result in acute radiation syndrome (ARS), which causes symptoms such as nausea, vomiting, diarrhea, and fatigue within hours of exposure. In severe cases, ARS can lead to organ failure and death within days or weeks. Long-term, even lower doses of radiation can significantly increase the risk of cancers, particularly thyroid cancer due to Iodine-131.

To test for dangerous levels of fallout in your immediate environment, it's essential to have access to radiation detection equipment, such as a Geiger counter or dosimeter. These tools measure radiation levels in the environment and help determine whether it's safe to go outside or consume food and water from potentially contaminated sources. A Geiger counter measures radiation in counts per minute (CPM) or micro Sieverts per hour (μSv/h), giving an indication of how much radiation is present in the environment.

Radiation isn't evenly distributed in fallout zones. In some areas, called "hot spots," radiation levels can be dangerously high, even in regions that are otherwise considered low-risk. This uneven distribution is a result of how fallout settles, influenced by factors such as wind patterns and terrain. Even within your shelter, you may find varying levels of radiation. This makes continuous monitoring with a Geiger counter essential for long-term survival.

Nuclear fallout is not a temporary problem that fades quickly. Depending on the scale of the nuclear event and the type of isotopes involved, fallout can linger for years, contaminating food, water, and soil. This presents unique challenges for long-term survival. Water sources, in particular, can remain contaminated for extended periods, requiring constant filtration and purification methods to ensure safety.

In summary, understanding nuclear fallout and its dangers is the first step in surviving a post-nuclear world. The science of radioactive decay, the health impacts of radiation exposure, and the importance of radiation detection all play vital roles in planning your survival strategy. Armed with this knowledge, you can make informed decisions about when to shelter, when to move, and how to protect yourself and your family from the invisible dangers of fallout.

The Fallout Map: Where Radiation Will Hit the Hardest

When a nuclear detonation occurs, the blast is just the beginning of the destruction. The real long-term threat comes from the fallout, which consists of radioactive particles that are carried into the atmosphere and then settle back to the ground over a wide area. Understanding where radiation will hit the hardest after a nuclear event is crucial for survival. This knowledge helps determine where to shelter, when it may be safe to move, and how to avoid the most dangerous regions.

Fallout distribution is influenced by several factors, including the size and type of the explosion, the altitude at which the detonation occurs, wind patterns, and weather conditions. While we can't predict exactly where fallout will land in every scenario, there are general principles and tools, such as fallout maps and simulations, that can give us a strong idea of where radiation will be most concentrated.

Types of Nuclear Detonations and Fallout

The type of nuclear detonation plays a significant role in how fallout spreads. A ground burst, where the bomb detonates on or near the surface of the Earth, generates much more fallout than an airburst, where the bomb detonates high above the ground. This is because a ground burst kicks up vast amounts of dirt, debris, and dust, which become irradiated and mix with the radioactive materials from the bomb itself. These particles then rise into the atmosphere and fall back down over time.

In contrast, an airburst, though devastating in terms of blast radius, generates less fallout because it doesn't suck up as much material from the ground. However, that doesn't mean an airburst is without fallout danger—it simply results in a different pattern of distribution.

The Fallout Cloud and Wind Patterns

After a detonation, the radioactive material forms a mushroom cloud, which rises high into the atmosphere. This cloud contains millions of microscopic particles of radioactive dust and debris. Depending on the altitude and the strength of the explosion, the fallout cloud can reach the stratosphere, where winds can carry it across vast distances.

Wind patterns play an essential role in where fallout will hit hardest. Local winds near the detonation site determine the initial direction of fallout, but high-altitude winds can transport radioactive particles much farther, sometimes across entire continents. This is why areas far from a detonation can still experience dangerous levels of radiation.

For instance, in the case of the Chernobyl disaster, radioactive fallout spread across Europe, reaching as far as Scandinavia and parts of the United Kingdom. A similar scenario could happen in the aftermath of a nuclear explosion, with fallout spreading hundreds or thousands of miles depending on wind direction and speed.

Fallout Zones: Near and Far

To better understand where radiation will hit the hardest, it's helpful to break the fallout map into different zones, based on proximity to the detonation site and the expected radiation levels.

Ground Zero (Primary Blast Zone): This is the area directly impacted by the explosion. In this zone, destruction is immediate and total. Radiation levels here will be lethally high, and survival is unlikely unless you are in a specially fortified underground shelter. Ground zero can extend several miles in every direction from the detonation site.

Severe Fallout Zone (Immediate Fallout Area): This area surrounds ground zero and can extend up to 10 to 20 miles from the blast site, depending on the size of the bomb. Fallout in this zone will be intense, with deadly levels of radiation present within the first few hours. Those who survive the blast but are exposed to fallout in this area face a high risk of acute radiation sickness (ARS), which can be fatal without immediate medical intervention.

Moderate Fallout Zone (Intermediate Fallout Area): Beyond the severe fallout zone is an area where radiation levels are lower, but still dangerous. This zone can extend anywhere from 50 to 100 miles from the detonation site, depending on wind patterns and terrain. Fallout here may not cause immediate death, but long-term exposure can lead to radiation sickness, cancers, and genetic damage. This zone is particularly dangerous because radiation may not be immediately noticeable, lulling survivors into a false sense of security.

Light Fallout Zone (Distant Fallout Area): Even hundreds of miles from the detonation, fallout can still pose a threat. This area will experience much lower radiation levels than the severe or moderate fallout zones, but it's important to note that radiation can still accumulate in food, water, and soil. In this zone, contamination of crops, livestock, and water sources becomes a serious concern, and long-term survival depends on proper testing and decontamination procedures.

Princeton University Simulation Studies

Princeton University has conducted extensive simulations on the impact of nuclear fallout, providing valuable insights into where radiation will hit the hardest in the event of a large-scale nuclear exchange. Their simulations show that major metropolitan areas and densely populated regions are the most vulnerable to both the immediate blast and the fallout that follows.

For example, in a simulated scenario of a nuclear detonation over a major city, the fallout cloud would not only affect the immediate area but would also drift downwind, contaminating rural areas hundreds of miles away. The simulation also reveals how different wind patterns can create unpredictable fallout maps, making it essential for survivors to stay informed and continually monitor radiation levels in their surroundings.

These studies suggest that while cities may be the primary targets of nuclear strikes, the surrounding areas—sometimes entire regions—can face just as much danger from fallout. Therefore, even those living in rural or suburban areas need to be prepared for the possibility of fallout reaching their homes.

Spotting Hot Spots on a Fallout Map

Fallout doesn't distribute evenly, and certain areas may experience higher concentrations of radiation than others. These are known as "hot spots." A hot spot can form when local wind patterns or terrain cause radioactive particles to settle in specific areas more heavily than others.

Mountains, valleys, and even buildings can affect the flow of air and, therefore, where fallout settles. For instance, wind moving through a mountain range might deposit higher levels of radioactive particles on one side of the range while sparing the other. Likewise, urban environments with tall buildings may create wind channels that direct fallout into certain neighborhoods while leaving others relatively unaffected.

Monitoring these hot spots is critical for long-term survival. Even if the general radiation levels in an area appear to be low, localized hot spots can still pose a serious risk. Geiger counters and other radiation detection tools are indispensable for identifying these zones.

Knowing Where to Shelter

Understanding where fallout will hit the hardest is not just about knowing where to avoid—it's also about knowing where to shelter effectively. In the event of a nuclear detonation, the most immediate concern is finding shelter that can protect you from both the blast and the fallout.

Basements and underground structures provide the best protection, as they can shield against both radiation and fallout particles. However, not all shelters are created equal, and those in the fallout zone will need to stay inside for an extended period, potentially weeks or even months, to avoid exposure to lethal levels of radiation.

In the aftermath of a nuclear detonation, knowing where radiation will hit the hardest can make the difference between life and death. Understanding the fallout map—how radiation spreads, where it accumulates, and how to avoid the most dangerous zones—gives you a strategic advantage in surviving a nuclear winter. Whether you're in an urban center, a rural area, or somewhere in between, staying informed, monitoring radiation levels, and knowing when and where to seek shelter are essential steps in surviving in a world polluted by fallout.

Radiation 101: Types, Effects, and Long-Term Threats

Radiation is a term that often evokes fear and uncertainty, especially when associated with nuclear fallout. However, understanding the different types of radiation, their effects on the human body, and the long-term threats they pose is crucial for survival in a post-nuclear world. In this chapter, we will break down the basics of radiation, explain how it interacts with living organisms, and discuss the potential long-term consequences for those exposed to it.

Types of Radiation

Radiation comes in different forms, each with its own properties and dangers. The types of radiation most relevant to nuclear fallout are ionizing radiation, which has enough energy to remove tightly bound electrons from atoms, thus creating charged particles called ions. There are three main types of ionizing radiation to be concerned with in a nuclear event:

Alpha Radiation: Alpha particles are relatively large and heavy compared to other types of radiation. They consist of two protons and two neutrons and are emitted by certain radioactive materials, such as Uranium-238 and Plutonium-239. While alpha particles are highly ionizing, meaning they can cause a lot of damage to living cells, they have very low penetration power. In fact, alpha particles can be stopped by a sheet of paper or even the outer layer of your skin. However, if alpha-emitting particles are inhaled or ingested, they can be extremely dangerous, as they can cause significant internal damage.

Beta Radiation: Beta particles are much smaller than alpha particles and consist of high-energy electrons or positrons. They are emitted by materials like Strontium-90 and Cesium-137, both of which are common byproducts of nuclear fission. Beta particles can penetrate the skin to a certain extent, potentially causing burns or other radiation damage to tissues. Like alpha particles, beta radiation becomes far more dangerous when radioactive particles are inhaled, ingested, or absorbed through the skin, as they can affect internal organs and tissues over time.

Gamma Radiation: Gamma rays are the most penetrating and dangerous type of radiation. They are not particles but high-energy electromagnetic waves, similar to X-rays but far more powerful. Gamma radiation is emitted by many radioactive isotopes, including Iodine-131, Cesium-137, and Cobalt-60. Unlike alpha and beta radiation, gamma rays can pass through the human body, penetrating deep into tissues and organs. This makes gamma radiation particularly hazardous, as it can cause widespread damage to cells and DNA. Protecting yourself from gamma radiation requires thick shielding, such as concrete, lead, or several feet of earth.

Effects of Radiation Exposure

Radiation exposure affects the human body in various ways, depending on the type, amount, and duration of exposure. The immediate effects of exposure to high levels of radiation are called acute radiation syndrome (ARS), commonly referred to as radiation sickness. ARS occurs when the body absorbs a large dose of radiation in a short period of time, usually from beta or gamma radiation.

Symptoms of ARS can begin within minutes to hours of exposure and include:

Nausea and vomiting

Fatigue

Diarrhoea

Skin burns or rashes

Hair loss

Bleeding or bruising

As ARS progresses, it can lead to more severe symptoms such as damage to internal organs, immune system suppression, and increased vulnerability to infections. If the exposure is high enough, ARS can be fatal within days or weeks.

However, even exposure to lower levels of radiation can be harmful, especially over long periods. Prolonged or repeated exposure to radiation, even at doses that don't cause immediate symptoms, can lead to chronic health problems such as:

Cancer: Radiation is a known carcinogen, meaning it can cause cancer by damaging DNA within cells. People exposed to radiation are at a higher risk of developing cancers, particularly leukemia and thyroid cancer. For example, Iodine-131, a common isotope found in nuclear fallout, is absorbed by the thyroid gland, significantly increasing the risk of thyroid cancer.

Genetic Mutations: Radiation can cause mutations in the DNA of reproductive cells, leading to genetic defects that can be passed on to future generations. These mutations may result in birth defects or an increased susceptibility to certain diseases.

Cardiovascular and Respiratory Issues: Long-term radiation exposure can increase the risk of heart disease, stroke, and lung problems, particularly if the lungs or heart were exposed to significant doses of radiation during the initial fallout period.

Cataracts: Radiation can damage the eyes, leading to cataracts and other vision problems over time.

Radiation Dose and Measurement

To understand how much radiation is dangerous, it's important to know how radiation exposure is measured. The most commonly used unit for measuring radiation dose is the sievert (Sv), which reflects the biological impact of the radiation on the human body. However, since even small doses of radiation can have significant effects, radiation doses are often measured in millisieverts (mSv) or microsieverts (µSv).

Here's a basic guide to radiation doses and their potential effects:

0 to 0.1 mSv (Normal background radiation): This is the amount of radiation you would be exposed to in a typical environment, such as from natural sources like cosmic rays and radon gas.

10 mSv (Medical diagnostic procedures): A single chest X-ray exposes a person to about 0.1 mSv, while a CT scan may expose someone to around 10 mSv.

50 to 100 mSv (Low-level exposure): While this level of exposure is considered relatively safe, it still increases the risk of long-term effects, such as cancer.

500 mSv (Moderate exposure): At this level, symptoms of radiation sickness may appear, especially after prolonged exposure. Medical intervention may be needed to mitigate damage.

1 to 2 Sv (Severe exposure): This level of exposure can cause serious symptoms of ARS, including nausea, fatigue, and an increased risk of cancer. Death is possible without immediate medical care.

4 to 6 Sv (Lethal exposure): Without immediate medical intervention, this level of exposure is usually fatal within weeks. Even with treatment, the survival rate is low.

10 Sv or higher (Fatal exposure): Death is almost certain within days or weeks due to severe damage to the internal organs and immune system.

Long-Term Threats of Radiation in a Post-Nuclear World

In the aftermath of a nuclear event, radiation poses long-term threats that may last for decades, even centuries, depending on the specific isotopes involved. One of the most concerning aspects of nuclear fallout is that radioactive isotopes like Cesium-137 and Strontium-90 have half-lives of approximately 30 years. This means they will remain hazardous for many years after the initial fallout.

The long-term threats from radiation include:

Contaminated Land and Water: Fallout can render vast areas of land and water sources unsafe for human habitation. Agricultural land may become unusable, and groundwater supplies may be contaminated with radioactive particles, making them unsafe to drink without proper filtration.

Food Supply Contamination: Crops, livestock, and wildlife may absorb radioactive materials, making food dangerous to consume. People living in affected areas may need to rely on stored food or grow their own using carefully controlled methods to avoid contamination.

Environmental and Ecological Damage: Radiation can devastate ecosystems, affecting plants, animals, and the delicate balance of food chains. Some species may become extinct or suffer from genetic mutations, leading to long-term ecological consequences.

Psychological and Social Impact: Living in a post-nuclear world also comes with significant psychological and social challenges. Survivors may experience trauma, anxiety, and depression due to the constant threat of radiation exposure and the breakdown of normal societal structures.

Radiation is an invisible but powerful force that can have devastating effects on human health and the environment. Understanding the types of radiation, their immediate and long-term effects, and the measurement of exposure levels is critical to surviving a nuclear event. In the aftermath of a nuclear explosion, the long-term threat of radiation will shape how people live, grow food, and interact with their environment for generations.

Simulating Catastrophe: Insights from Princeton University Nuclear Studies

Princeton University has long been at the forefront of nuclear research, using cutting-edge simulations to model the devastating impacts of nuclear warfare. These simulations, often conducted as part of the university's Program on Science and Global Security, provide a grim but necessary look into the potential consequences of nuclear conflict. By using highly sophisticated modeling software and powerful computers, Princeton's researchers have been able to create detailed scenarios of how a nuclear war might unfold, from the immediate destruction of cities to the long-term effects of nuclear fallout. In this chapter, we'll explore some of the key insights from these simulations and what they reveal about surviving a nuclear catastrophe.

The Princeton Nuclear War Simulation

In one of the most widely referenced studies, Princeton University created a simulation that explored the consequences of a nuclear conflict between NATO and Russia. This scenario involved the use of hundreds of nuclear warheads, aimed at both military and civilian targets across Europe and the United States. The results were chilling. In the first few hours alone, the simulation predicted that over 90 million people would be killed or injured, with many more affected by the subsequent fallout.

The simulation's main focus was to highlight the rapid escalation of such a conflict, beginning with a single tactical nuclear strike and culminating in an all-out nuclear war involving thousands of warheads. Once this level of destruction begins, there's little time to react. Within minutes, entire cities would be obliterated, with survivors facing the grim reality of radioactive fallout and the collapse of societal structures.

While this particular scenario involved large-scale global conflict, the insights gained from the simulation apply to any nuclear event, whether it's a limited exchange or a full-scale war. Understanding how fallout spreads, where the highest levels of radiation will be concentrated, and what the long-term environmental and health effects might be are crucial for anyone preparing to survive a nuclear winter.

Fallout Patterns: How Radiation Travels

One of the most important insights from the Princeton simulation is how nuclear fallout spreads following a detonation. As discussed in earlier chapters, fallout consists of radioactive particles that are lifted into the atmosphere by the explosion and then carried by wind and weather patterns, falling back to Earth over the course of days, weeks, or even longer.

Princeton's simulation showed that fallout can travel much farther than many people realize. In the NATO-Russia scenario, for example, radioactive clouds drifted across Europe, covering large parts of the continent in dangerous levels of radiation. Fallout from strikes in the U.S. spread across North America, affecting not just the immediate areas around the explosion sites, but entire regions far from the initial blasts.

This means that even if you're not located near a primary target—such as a major city or military installation—you could still be at risk from fallout. The simulation highlighted that areas hundreds, even thousands, of miles away from the detonation sites could experience lethal doses of radiation if the wind patterns carry fallout in their direction. This makes it critical for survivalists to understand not just their proximity to potential nuclear targets, but also the prevailing wind and weather patterns that could influence fallout distribution.

Time Is of the Essence: The Importance of Immediate Action

Another key takeaway from Princeton's nuclear simulations is the need for immediate action in the event of a nuclear strike. Once a detonation occurs, there is a very short window of time to seek shelter before the fallout begins to settle. Fallout can start to descend within 15 minutes to an hour after the explosion, depending on factors like the altitude of the detonation and local weather conditions.

The simulation revealed that those who take shelter immediately, especially in areas shielded from radioactive particles like basements or underground bunkers, have a significantly higher chance of survival than those who delay. Every second counts in the aftermath of a nuclear strike, and knowing where and how to seek shelter can make the difference between life and death.

Princeton's research also emphasized the importance of staying sheltered for an extended period. Radiation levels are highest in the first few hours and days after fallout begins, but they gradually decrease over time. However, the simulation showed that it can take weeks, even months, for radiation levels to drop to relatively safe levels, depending on the size of the nuclear event and the types of radioactive isotopes involved. This means that survivors may need to stay in their shelters for long periods, requiring them to be prepared with enough food, water, and supplies to last for weeks on end.

Hotspots and Safe Zones: Identifying Areas of Risk

One of the more nuanced findings from Princeton's simulations was the identification of radiation "hotspots" and areas of relative safety. As fallout spreads, it doesn't do so evenly. Instead, wind patterns, terrain, and even buildings can cause fallout to accumulate in certain areas more heavily than others. These hotspots may have radiation levels that are several times higher than surrounding areas, posing an increased risk to anyone living or traveling through them.

The simulation also identified what could be considered "safe zones"—areas that, due to wind patterns or geographical features, experience significantly lower levels of fallout. In some cases, valleys or other natural features shielded parts of the landscape from the worst of the fallout, creating potential refuge areas for survivors. However, these safe zones can be temporary, as shifting winds or further nuclear detonations may change the distribution of fallout over time.

For preppers and survivalists, understanding how to identify both hotspots and safe zones is critical. Relying on maps, weather reports, and radiation detection equipment like Geiger counters can help determine where to take shelter and when it's safe to move. Princeton's simulation underscores the importance of being mobile and adaptable, as staying in one location for too long could expose you to shifting radiation patterns.

Long-Term Environmental and Health Effects

The Princeton simulations didn't just focus on the immediate aftermath of a nuclear war—they also explored the long-term consequences of radiation exposure and environmental contamination. Even after the initial fallout has settled, radioactive isotopes like Cesium-137 and Strontium-90 can remain in the environment for decades, poisoning water sources, soil, and food supplies.

The simulation showed that areas affected by heavy fallout may be uninhabitable for years, or even decades, after a nuclear event. Crops grown in contaminated soil can absorb radioactive particles, making them unsafe to eat. Livestock and wildlife may also become contaminated, adding another layer of danger to the post-nuclear landscape.

Human health is similarly affected by long-term radiation exposure. Survivors may face increased risks of cancer, genetic mutations, and other chronic health problems due to their exposure to fallout. Children and pregnant women are particularly vulnerable, as their developing bodies are more sensitive to radiation.

In the long term, surviving a nuclear winter means more than just avoiding the initial fallout—it requires a comprehensive strategy for avoiding contaminated areas, securing clean water and food, and monitoring radiation levels in the environment.

The Role of International Response

One of the more sobering insights from Princeton's nuclear studies is the limited role that international response can play in the immediate aftermath of a nuclear exchange. In large-scale scenarios like the NATO-Russia simulation, infrastructure is so severely damaged that traditional emergency services and governmental support may be unavailable for weeks or even months. Hospitals and medical facilities may be overwhelmed, or even destroyed, leaving survivors to rely on their own resources and preparation.

This insight stresses the importance of individual preparedness. Preppers and survivalists need to be self-sufficient, not just in terms of food and water, but also in terms of medical supplies, radiation detection equipment, and alternative sources of communication. In the absence of government aid, local communities may need to band together to form their own support systems, trading resources and skills to ensure mutual survival.

Princeton University's nuclear simulations offer invaluable insights into the catastrophic consequences of nuclear warfare. From understanding how fallout spreads to identifying radiation hotspots and safe zones, these simulations provide critical information for anyone preparing to survive a nuclear event. The simulations also highlight the importance of immediate action, long-term planning, and self-sufficiency in a world where traditional infrastructures may no longer exist.

By learning from these simulations, preppers and survivalists can better anticipate the challenges they will face and develop strategies to protect themselves and their loved ones in the event of a nuclear catastrophe.

Testing for Radiation: Tools and Techniques

Testing for radiation is one of the most critical survival skills in a post-nuclear world. Understanding the tools and techniques needed to measure radiation levels can be the difference between life and death, as radiation is invisible, odourless, and undetectable without specialized equipment. In this chapter, we'll explore the various types of radiation detection tools, how they work, and the best practices for testing and monitoring radiation in the environment, on food and water, and even on people.

Why Testing for Radiation is Essential

Radiation is not evenly distributed after a nuclear event. Fallout patterns depend on many variables, such as wind direction, terrain, and the altitude of the nuclear explosion. As a result, some areas may be heavily contaminated while others, even nearby, may have relatively low levels of radiation. Without proper testing, it's impossible to know whether an area is safe for habitation, whether food and water are safe for consumption, or whether people have been exposed to harmful levels of radiation.

Even after the initial fallout has settled, radioactive particles can linger in the environment for months or years, making regular testing a necessity for long-term survival. Additionally, "hot spots" of concentrated radiation can form, especially in low-lying areas or where fallout accumulates due to wind patterns. Knowing how to test for radiation will help you avoid these dangerous zones and keep your family safe.

Radiation Detection Tools

There are several types of radiation detection tools available, each suited for different tasks. These tools measure radiation in different ways and have varying levels of sensitivity, making some more appropriate for personal use and others better suited for environmental testing.

Geiger Counter (Geiger-Müller Detector)

The Geiger counter is one of the most commonly used tools for detecting radiation and is often the go-to device for preppers and survivalists. It works by using a Geiger-Müller tube, which becomes ionized when exposed to radiation. This ionization generates an electric pulse that the Geiger counter translates into a reading, typically displayed as counts per minute (CPM) or microsieverts per hour (μSv/h), which measures the rate of radiation exposure.

Geiger counters are particularly good for detecting beta and gamma radiation, though some models can also detect alpha radiation. They are portable and easy to use, making them ideal for quickly checking the radiation levels in a given area, on objects, or on people.

When using a Geiger counter:

Ensure proper calibration: Many Geiger counters require regular calibration to maintain accuracy. Be sure to check the device's manual and calibrate it according to the manufacturer's guidelines.

Test different surfaces: Fallout may settle unevenly, so it's important to test multiple areas, such as the ground, walls, or objects you may come into contact with.

Monitor regularly: Radiation levels can fluctuate based on wind patterns, weather, and the movement of radioactive particles, so it's important to test frequently, especially if you're near a known fallout zone.

Dosimeter

A dosimeter is a device that measures cumulative radiation exposure over time, rather than detecting radiation levels at a specific moment like a Geiger counter. Dosimeters are typically worn on the body, making them useful for monitoring how much radiation a person has been exposed to over days, weeks, or even months.

There are different types of dosimeters:

Electronic dosimeters: These devices provide real-time readings and can be reset after each use. They are ideal for continuous monitoring in environments where radiation exposure is a concern.

Film badge dosimeters: These are less common among survivalists but are still used in some industries. They work by exposing a small piece of film to radiation, which can then be developed to determine the total radiation dose received.

Dosimeters are especially useful for long-term exposure monitoring, helping you track your cumulative radiation dose to avoid reaching dangerous levels over time. If you plan on venturing into areas where fallout is present, wearing a dosimeter is a must.

Scintillation Detector

A scintillation detector is a more sensitive device than a Geiger counter and is capable of detecting even low levels of radiation. It works by using a special material that emits light (scintillates) when struck by radiation. This light is then converted into an electrical signal, which can be measured to determine the level of radiation.

Scintillation detectors are particularly useful for detecting alpha particles, which Geiger counters may struggle with. They are also highly sensitive to gamma radiation, making them ideal for more precise measurements in areas where even small amounts of radiation could pose a long-term threat.

These detectors are more expensive and less portable than Geiger counters, so they are often used in conjunction with other tools when more accurate readings are needed, such as testing food and water for contamination.

Personal Radiation Alert Devices

For those who want a more passive form of radiation detection, personal radiation alert devices (often called dosimeter alarms or radiation pagers) provide a compact, portable way to detect when radiation levels are dangerously high. These devices are typically worn on the body and will emit an audible alarm or vibration if radiation levels exceed a predetermined threshold.

These devices don't provide detailed readings like a Geiger counter but are useful for providing a quick warning if you accidentally enter a radioactive area or if radiation levels suddenly spike in your environment.

Techniques for Testing Radiation

Using radiation detection tools effectively involves understanding how to test different areas, surfaces, and materials for contamination. The techniques you use can vary depending on whether you're testing the environment, food and water, or people.

Testing the Environment

When testing outdoor areas, such as your shelter's surroundings or potential evacuation routes, it's essential to use a Geiger counter or scintillation detector to scan the ground, walls, and other surfaces. Fallout tends to accumulate more heavily in low-lying areas, so pay special attention to valleys, drainage ditches, and other depressions.

Ground testing: Start by holding the Geiger counter close to the ground, moving slowly across the area you want to test. It's important to scan multiple locations, as fallout may not settle evenly. Focus on areas where dust and debris may have collected, such as gutters, window sills, or doorways.

Air testing: While less common, some areas may still have radioactive particles suspended in the air, particularly after recent fallout. If you have an air filter with a HEPA setting, you can run it in the area and then test the filter with your Geiger counter to determine whether radioactive particles are present in the air.

Testing Food and Water

Fallout can contaminate food and water sources, making it crucial to test anything you plan to consume. For food, use a Geiger counter or scintillation detector to scan the surface. If the food was grown in contaminated soil or exposed to fallout, radioactive particles could be present on the surface or absorbed into the plant or animal tissues.

Testing canned food: Canned goods are one of the safest food sources after a nuclear event, as they are sealed and protected from fallout. However, it's still wise to test the outside of cans for contamination before handling them.

Water testing: Use a scintillation detector or water testing kit designed for radiation detection to test water sources. Filtered water may still contain dissolved radioactive isotopes, so testing even filtered water is necessary.

Testing People

After a nuclear fallout event, it's important to regularly check people for radioactive contamination, especially if they've been outside or exposed to dust and debris. Use a Geiger counter to scan clothing, skin, and hair for radioactive particles.

Clothing: Fallout can cling to clothing, so it's essential to change out of contaminated clothes and wash them separately, if possible. A Geiger counter can help identify whether radioactive particles are still present on the fabric.

Skin and hair: Pay close attention to exposed skin and hair. If contamination is found, decontamination involves washing thoroughly with soap and water, being careful not to scrub too hard, which can open the skin and allow radioactive particles to enter the body.

Best Practices for Testing Radiation

Test regularly: Radiation levels can change over time, so frequent testing is essential, especially if you live in or are traveling through an area affected by fallout.

Be consistent: Develop a routine for testing key areas, food, water, and people. Consistency ensures that you catch any changes in radiation levels as soon as they happen.

Stay informed: Pay attention to reports from official sources about radiation levels in your region, but always verify those readings with your own testing tools. Even in low-risk areas, localized hotspots can develop unexpectedly.

Testing for radiation is a vital skill for anyone trying to survive in a post-nuclear world. The tools and techniques discussed in this chapter provide a comprehensive approach to detecting and monitoring radiation levels, ensuring that you can avoid hazardous areas, protect your food and water supplies, and decontaminate yourself and others if necessary. By regularly testing and staying vigilant, you'll be better equipped to navigate the long-term challenges posed by nuclear fallout and radiation exposure.

The Geiger Counter: Your First Line of Defense

The Geiger counter is perhaps the most essential tool for survival in a post-nuclear environment. As the first line of defense against invisible and deadly radiation, this device allows you to detect, measure, and monitor radiation levels in the air, on surfaces, in food and water, and even on people. Understanding how to properly use and interpret a Geiger counter's readings is a critical survival skill, particularly in a world contaminated by nuclear fallout. In this chapter, we'll take a deep dive into how the Geiger counter works, how to use it effectively, and why it's indispensable for anyone navigating a nuclear winter.

How the Geiger Counter Works

At its core, the Geiger counter detects ionizing radiation—radiation that has enough energy to knock electrons off atoms, creating charged particles. It works using a Geiger-Müller (GM) tube, which is filled with a gas (usually a mix of inert gases like argon or helium) that becomes ionized when struck by radioactive particles. When ionizing radiation, such as alpha, beta, or gamma particles, enters the tube, it causes the gas inside to ionize, creating an electrical charge. This charge generates an electrical pulse, which the Geiger counter counts and then translates into a reading.

The Geiger counter typically measures radiation levels in one of two ways:

Counts per minute (CPM) or counts per second (CPS): This measures the number of radioactive particles detected by the GM tube over time. Higher CPM or CPS readings indicate higher levels of radiation.

Microsieverts per hour (μSv/h) or millisieverts per hour (mSv/h): These units measure the dose of radiation exposure over time. They give you an indication of how much radiation is present in the environment and how much you are likely to absorb if you remain in that area.

A Geiger counter emits a clicking sound when it detects radiation, with faster clicks indicating higher levels of radiation. This audible feature makes it easy to assess radiation levels even without closely watching the screen.

Types of Radiation Detected by a Geiger Counter

A standard Geiger counter is capable of detecting three types of ionizing radiation: alpha, beta, and gamma. However, it's important to understand that different types of radiation require different levels of protection, and some may be more dangerous than others.

Alpha particles are large and slow-moving. They can't penetrate the skin and are generally harmless unless inhaled or ingested. While many Geiger counters can detect alpha radiation, they may require a special probe to do so effectively.

Beta particles are smaller and faster. They can penetrate the outer layers of skin and cause radiation burns, making them more dangerous. A Geiger counter is usually effective in detecting beta radiation, although the device may need to be close to the source.

Gamma rays are the most dangerous form of radiation because they can penetrate deeply into tissues and organs, causing severe internal damage. Geiger counters are highly sensitive to gamma radiation, which is why they are so vital in a nuclear fallout scenario.

Understanding which type of radiation you're dealing with can help you determine the appropriate level of protective measures. For example, gamma radiation requires thick shielding, while alpha particles can be blocked by something as simple as a sheet of paper or protective clothing.

How to Use a Geiger Counter Effectively

The Geiger counter is a powerful tool, but only if used correctly. Whether you're testing the environment around you, checking your food and water supplies, or monitoring people for radiation exposure, there are key practices you need to follow.

Testing the Environment

In the aftermath of a nuclear event, testing your immediate environment is crucial. Radioactive fallout can settle on buildings, streets, vegetation, and in the soil. Here's how to test the environment around you:

Move slowly and systematically: Hold the Geiger counter a few inches above the surface you are testing and move slowly across the area. Because radiation can settle unevenly, especially in "hot spots," it's important to scan multiple locations to get an accurate reading.

Check key points: Fallout tends to accumulate in low-lying areas like ditches, gutters, and depressions in the ground. Focus on testing these areas first, as they are more likely to have higher concentrations of radioactive particles.

Test frequently: Radiation levels can fluctuate based on wind patterns, weather changes, or further nuclear detonations. Make it a habit to test your environment regularly to ensure that conditions haven't worsened unexpectedly.

Testing Food and Water

Contaminated food and water are some of the greatest hazards in a fallout scenario. While it's safer to rely on stored or sealed food and water supplies, there may come a time when you need to test these resources for radioactive contamination.

Food testing: Hold the Geiger counter close to the surface of food items, especially those that were exposed to the air after fallout occurred. Freshly grown produce and animal products are most at risk of contamination. Even if the food is canned or packaged, you should still test the outside packaging to ensure it's safe to handle.

Water testing: Testing water sources can be tricky, as radiation in water may be dissolved rather than particulate. If you have access to a scintillation detector, it's ideal for water testing. However, if you're relying on a Geiger counter, you can test by running water through a filter (such as a HEPA filter) and testing the residue collected in the filter.

Testing People for Radiation

After being exposed to fallout, it's essential to check yourself and others for radioactive contamination. Fallout particles can settle on clothing, skin, and hair, posing a significant risk if not properly cleaned.

Scan the body: Use the Geiger counter to scan the entire body, starting with areas most exposed to the air, like the hands, face, and head. Pay special attention to hair, as radioactive dust can cling to it.

Check clothing: Clothing can trap radioactive particles, so it's important to test it before taking it off or handling it. If contamination is found, clothes should be removed carefully and disposed of in a safe area. If possible, contaminated clothing should be washed separately from other items.

Testing frequently: If you've been outside or near potential fallout zones, make it a routine to test for radiation when you return indoors. Even if the initial exposure was low, radioactive particles can accumulate over time.

Interpreting Geiger Counter Readings

Understanding what your Geiger counter's readings mean is just as important as using the tool itself. The readings will help you determine whether an area is safe, whether food and water are contaminated, or if someone needs decontamination.

0.05–0.20 μSv/h: Normal background radiation. This is the level of radiation you would expect in most areas under normal circumstances. These readings are generally safe.

0.20–1.00 μSv/h: Elevated radiation. This range may be higher than normal but generally doesn't pose an immediate health risk. Prolonged exposure, however, can lead to increased risks of cancer or other long-term health effects.

1.00–10.00 μSv/h: Dangerously high radiation. Prolonged exposure to these levels can lead to radiation sickness and long-term health problems. Immediate action should be taken to seek shelter or leave the area.

10.00 μSv/h and above: Critical radiation levels. At these levels, immediate exposure can cause acute radiation syndrome (ARS), which can be fatal if not treated. Leave the area immediately and seek shelter in a well-shielded location.

Maintaining Your Geiger Counter

A Geiger counter is only useful if it's well-maintained. Like any tool, it needs regular care and attention to ensure accurate readings.

Battery check: Many Geiger counters run on batteries, and it's crucial to ensure that your device has a reliable power source. Always carry spare batteries, as your Geiger counter may be your only warning system in a high-radiation zone.

Calibration: Some Geiger counters require regular calibration to ensure they remain accurate. Check the manufacturer's instructions to determine whether and how often your device needs to be calibrated.

Protect the sensor: The GM tube inside your Geiger counter is delicate and can be damaged by physical shocks or exposure to moisture. Handle the device with care and store it in a protective case when not in use.

Measuring Danger: How to Interpret Radiation Levels

Interpreting radiation levels accurately is critical to making informed decisions about your safety in a post-nuclear world. Radiation is invisible and odorless, so without a reliable way to measure and understand it, you could unknowingly expose yourself to dangerous levels. In this chapter, we'll explore how to interpret radiation levels using common measurement units, what different levels of radiation exposure mean for your health, and how to make decisions based on the readings from your Geiger counter or other detection tools.

Units of Radiation Measurement

Radiation is measured in several units, each reflecting different aspects of radiation exposure. Understanding these units is the first step to interpreting radiation levels accurately.

Sieverts (Sv) and Millisieverts (mSv)

The sievert (Sv) is the unit used to measure the biological effect of ionizing radiation on the human body. It accounts for the type of radiation and its impact on living tissue. However, because 1 sievert represents a very high level of radiation exposure, radiation is typically measured in smaller units: millisieverts (mSv), which are one-thousandth of a sievert, and microsieverts (μSv), which are one-millionth of a sievert.

1 Sv = 1,000 mSv

1 mSv = 1,000 μSv

When using a Geiger counter, the most common unit displayed is microsieverts per hour (μSv/h), which tells you how much radiation you are being exposed to per hour in that specific location.

Grays (Gy)

The gray (Gy) measures the absorbed dose of radiation, or the amount of energy deposited in a material, typically human tissue. While sieverts account for the biological effects of radiation, grays are focused solely on the physical absorption. The gray is often used in medical and scientific contexts but is less common in survival scenarios compared to sieverts.

Counts per Minute (CPM) or Counts per Second (CPS)

CPM and CPS measure the number of radioactive particles detected by a Geiger counter over a specific period. While these units don't directly translate into the biological effect of radiation (like sieverts), they can give you a rough idea of how much radiation is present. Different Geiger counters have different sensitivity levels, so the same radiation level might result in different CPM readings depending on the device.

CPM is often used when testing specific objects or food to detect contamination.

CPS is a similar unit but counts particles on a second-by-second basis.

Interpreting Common Radiation Levels

To make informed decisions in a radiation-affected environment, it's essential to understand what different levels of radiation mean for your health and safety. Below are some commonly encountered radiation levels and their potential effects:

0.05 to 0.2 µSv/h (Normal Background Radiation)

This is the normal range of background radiation found in most parts of the world. It comes from natural sources like cosmic rays, radon gas, and even some building materials. This level of radiation is harmless and does not pose any health risks.

0.2 to 1 µSv/h (Slightly Elevated Radiation)

Slightly elevated levels of radiation can be found in areas with higher-than-average background radiation due to geological factors, or in areas affected by low-level fallout. While this level isn't immediately dangerous, long-term exposure may slightly increase the risk of cancer over time. Monitoring these levels is important, but it's generally safe to remain in the area for short periods.

1 to 10 µSv/h (Dangerous Over Time)

Radiation levels in this range are cause for concern. Although short-term exposure (a few hours) may not lead to immediate health effects, staying in an area with this level of radiation for extended periods (weeks or months) can lead to an increased risk of cancer or other long-term health effects. It is important to minimize time spent in such areas and to seek better-shielded shelter if possible.

10 to 100 µSv/h (Potential Health Risk)

Prolonged exposure to radiation levels in this range can cause noticeable health issues. Spending hours or days in this environment could lead to symptoms of radiation sickness, such as nausea, fatigue, and weakened immune function. Evacuating or finding a well-shielded shelter is strongly recommended.

100 to 1,000 µSv/h (Acute Radiation Exposure)

At this level, exposure becomes increasingly hazardous. Short-term exposure can cause significant radiation sickness, and prolonged exposure can be fatal. Even spending a few hours in such an environment could lead to serious health problems. Evacuation or sheltering in a fortified, shielded area (such as an underground bunker) is critical.

1,000 to 10,000 µSv/h (Lethal Levels of Radiation)

Exposure to radiation levels in this range is life-threatening. Immediate symptoms of radiation sickness, such as vomiting, diarrhea, and burns, will occur, and long-term survival is unlikely without medical intervention. Even short-term exposure (minutes to hours) can be fatal. Evacuating from the area or moving to the deepest, most shielded shelter available is essential.

10,000 µSv/h and Above (Instantly Lethal Radiation)

Radiation at this level will cause severe damage to the body in a matter of minutes. Death is highly likely within hours or days. Survivors of such exposure will require immediate and advanced medical care to have any chance of survival. Avoiding these areas entirely is crucial for survival.

Practical Use of Radiation Levels for Decision-Making

Knowing how to interpret radiation levels is one thing, but applying that knowledge to real-world situations is what ensures survival. Here are some guidelines for making decisions based on your Geiger counter readings:

When to Shelter

If you detect radiation levels above 10 µSv/h, it's time to shelter. The higher the level, the more protection you will need. Basements or underground shelters provide better protection from gamma radiation, which penetrates most materials. Shielding yourself with concrete, lead, or several feet of packed earth can significantly reduce exposure.

Even if the radiation level is not immediately life-threatening, sheltering for several days can protect you from higher doses in the long term. Radiation levels tend to decrease as time passes after a nuclear event, so staying indoors during the first few days is crucial.

When to Evacuate

If radiation levels exceed 100 µSv/h and show no signs of dropping, evacuation may be necessary. It's important to have an evacuation plan that takes into account potential fallout patterns and wind directions. When evacuating, minimize exposure by traveling quickly and covering exposed skin with protective clothing, and ensure you carry a Geiger counter to monitor radiation levels along your route.

If you must pass through higher radiation zones, try to do so quickly and efficiently. Always have an alternative route in case your primary path becomes too dangerous.

When It's Safe to Go Outside

After a nuclear event, one of the most critical decisions is knowing when it's safe to venture outside. Radiation levels tend to decrease over time, especially during the first few days and weeks after fallout. As a general rule:

If radiation levels are below 1 µSv/h, it is relatively safe to go outside for short periods, though long-term exposure should still be minimized.

If levels are between 1 and 10 µSv/h, it's best to limit your time outside to essential tasks only, such as gathering food or checking on your shelter.

If levels are above 10 µSv/h, remain sheltered as much as possible and only venture out if absolutely necessary.

Testing Food and Water

Testing food and water is crucial when radiation levels in the environment are elevated. Even if the air around you shows safe readings, fallout can contaminate crops, livestock, and water sources. Use your Geiger counter to test food and water before consumption, especially if it comes from an outdoor source exposed to fallout.

Radiation levels around food should ideally be below 0.2 µSv/h. If levels are higher, it's best to avoid consumption unless it's an emergency and no other food source is available.

Long-Term Health Considerations

Even after immediate radiation exposure is reduced, long-term health risks remain, particularly in areas that have been contaminated with fallout. Continuous low-level exposure can lead to increased risks of cancers, particularly thyroid cancer due to Iodine-131, as well as genetic damage, birth defects, and cardiovascular problems. For these reasons, ongoing monitoring of radiation levels in both the environment and food sources is essential for long-term survival.

If you are exposed to radiation over extended periods, it's important to track your cumulative dose using a dosimeter or similar device. This will help you assess whether you've received a potentially harmful amount of radiation and whether medical intervention is necessary.

Understanding how to measure and interpret radiation levels is a fundamental survival skill in a nuclear fallout scenario. By knowing what different levels of radiation mean and how to use tools like a Geiger counter effectively, you can make informed decisions about when to shelter, when to evacuate, and when it's safe to go outside. Interpreting these levels accurately will help you navigate the dangers of nuclear fallout and protect yourself and your loved ones from the invisible threat of radiation.

Decontaminating People after Exposure

Decontaminating people after exposure to nuclear fallout is one of the most urgent tasks in a post-nuclear scenario. Fallout consists of radioactive particles that can cling to clothing, skin, and hair, continuing to emit dangerous radiation long after the initial explosion. Immediate and effective decontamination is critical to reducing the risk of radiation poisoning and limiting further exposure to those around you. In this chapter, we'll discuss the step-by-step process for decontaminating individuals after exposure, focusing on how to safely remove radioactive particles from clothing, skin, and hair, as well as best practices to follow to ensure long-term health.

Why Decontamination Is Critical

Radioactive fallout consists of tiny particles of dust and debris that become radioactive after being propelled into the atmosphere by a nuclear explosion. These particles can settle on anything they touch, including buildings, vehicles, vegetation, and, most critically, people. When fallout lands on a person, it poses a direct radiation hazard to their health, particularly if they are exposed for an extended period.

Radiation exposure through fallout can occur in three main ways:

External exposure: Fallout on the skin, clothing, and hair can emit radiation that penetrates the body.

Inhalation: Breathing in radioactive particles can lead to internal radiation exposure, affecting the lungs and other internal organs.

Ingestion: Consuming contaminated food or water can introduce radioactive particles into the digestive system.

Decontaminating individuals as soon as possible after exposure is essential to minimizing the health risks associated with radiation. Removing radioactive particles quickly reduces the total dose a person receives and prevents those particles from being inhaled or ingested. It also helps to prevent the spread of contamination to others or to safe areas.

Immediate Decontamination: The First Steps

The first priority after a person has been exposed to fallout is to decontaminate them as quickly as possible to limit the amount of radiation they absorb. If decontamination supplies are available, they should be used immediately, but even without specialized equipment, there are ways to significantly reduce the risk from fallout.

Move to a Safe Area

The first step in decontamination is to move away from the contaminated area. Fallout particles will continue to rain down in the hours after a nuclear explosion, so it's important to seek shelter as quickly as possible, ideally in an enclosed space with thick walls, such as a basement, bunker, or interior room of a building.

If possible, avoid walking through areas that are heavily contaminated, as this will only increase your exposure. Once you are in a safer environment, you can begin the decontamination process.

Remove Contaminated Clothing

One of the fastest and most effective ways to reduce radiation exposure is to remove contaminated clothing. Studies have shown that simply removing outer clothing can eliminate up to 90% of radioactive particles from the body. Here's how to do it safely:

Be cautious: When removing clothing, be careful not to shake it or disturb the particles, as this can cause them to become airborne and increase the risk of inhalation.

Remove outer layers first: Start with jackets, hats, and shoes—anything that has been directly exposed to the outside environment. Place contaminated clothing in a sealed plastic bag or container to prevent the spread of radioactive particles.

Avoid touching your face: Try not to touch your face or any other exposed skin while removing clothing, as this could transfer radioactive particles to your skin or mouth.

If water is available, you can rinse the clothing before removing it, but the priority is getting it off your body as quickly as possible.

Bag and Dispose of Contaminated Materials

Once clothing and other contaminated materials have been removed, they should be sealed in plastic bags or containers to prevent further spread of fallout. If you have access to a Geiger counter, test the bag or container to confirm that the radioactive particles are contained.

Dispose of these materials as far away from living and shelter areas as possible. If possible, bury the contaminated items in the ground to shield others from the lingering radiation.

Washing Off Fallout: Decontaminating Skin and Hair

After removing contaminated clothing, the next step is to wash off any radioactive particles that may have settled on the skin or hair. This process should be done with great care, as scrubbing too hard or using the wrong cleaning methods could actually worsen the situation by opening the skin to contamination or spreading fallout particles further.

Shower with Soap and Water

The most effective way to decontaminate skin and hair is to take a shower using lukewarm water and mild soap. Follow these steps to ensure the most thorough decontamination:

Use lukewarm water: Hot water can open the pores of your skin, increasing the risk of radioactive particles being absorbed. Cold water, on the other hand, may not effectively remove fallout particles. Lukewarm water is ideal for washing fallout off your skin.

Avoid harsh scrubbing: Gently wash your skin with soap, paying special attention to areas that were most exposed, such as the hands, face, and hair. Do not scrub too hard, as this could cause skin irritation or open cuts that allow radioactive particles to enter the bloodstream.

Wash hair carefully: Use shampoo, but avoid using conditioner, as conditioner can cause radioactive particles to bind to your hair, making them harder to remove.

Rinse thoroughly: Make sure to rinse your body and hair thoroughly to ensure that all particles are washed away. Focus on areas where fallout may have accumulated, such as the neck, ears, and under fingernails.

If running water is not available, you can use a wet cloth or sponge to wipe down the skin, but this should only be a temporary solution until you can access a proper shower.

Avoid Contaminating Clean Areas

When washing off fallout, it's crucial to avoid contaminating clean areas. Here are some best practices:

Contain runoff water: If possible, contain the water used for decontamination in a specific area and dispose of it far from living spaces. If you're in an outdoor shelter, direct runoff water away from your shelter and food supplies.

Wear gloves if available: If assisting someone else with decontamination, wear gloves to prevent spreading radioactive particles from the person's skin or clothing to your own.

Minimize contact with contaminated surfaces: Be mindful of any surfaces you touch after decontamination, as radioactive particles can transfer to clean areas.

Addressing Cuts and Abrasions

If a person has cuts, abrasions, or other open wounds, it's critical to clean them carefully without allowing radioactive particles to enter the bloodstream. Here's how to handle injuries:

Cover open wounds: Before starting decontamination, cover any open wounds with clean, dry cloth or gauze to prevent radioactive particles from entering. If you must clean an open wound, do so with clean water or an antiseptic solution.

Wash around wounds: Be extra gentle when washing around injured areas, and avoid scrubbing or rubbing the skin.

Once decontamination is complete, reapply clean bandages and monitor the wound for any signs of infection or irritation.

Monitoring for Residual Contamination

After decontaminating, it's important to monitor for any residual radiation. Using a Geiger counter or other radiation detection device, test the individual's skin, clothing, and any objects they handled during the decontamination process. Focus on high-risk areas, such as:

Hands and wrists

Face, neck, and ears

Hair and scalp

Feet and shoes

If radiation levels are still detectable, additional washing or decontamination may be required. Keep testing until radiation levels fall within safe limits, as discussed in previous chapters.

Decontamination in the Absence of Water

In some cases, running water may not be available for decontamination. If this is the case, there are alternative methods for removing radioactive particles:

Use wet wipes or damp cloths: If clean water is unavailable, wet wipes or damp cloths can be used to gently wipe down the skin. Focus on exposed areas, especially the hands, face, and hair.

Remove as much dust as possible: If dry conditions make it impossible to use water or wipes, brush off as much fallout dust as possible using a clean cloth or soft brush. Be sure to wash off fully as soon as water becomes available.

Cover exposed skin: In the absence of proper decontamination facilities, covering as much of your skin as possible with clean clothing, masks, or cloths will help minimize further exposure.

Preventing Recontamination

Decontamination is an ongoing process in a fallout environment. Even after a thorough washing, it's important to take steps to prevent recontamination. Here's how to stay safe:

Wear protective clothing: If you need to venture outside again, wear protective clothing, including a hat, long sleeves, gloves, and a mask or scarf to cover your mouth and nose.

Minimize exposure: Limit your time outside, especially in areas where fallout is known to be heavy. The less time you spend exposed to fallout, the lower your risk of contamination.

Test regularly: Use a Geiger counter to monitor radiation levels in your environment and on your clothing to ensure you are staying within safe limits.

Cleaning Tainted Food: What You Need to Know

In the aftermath of a nuclear event, one of the most pressing challenges for survival is ensuring that the food you eat is free from radioactive contamination. Fallout can settle on crops, livestock, and stored food supplies, making it a potential hazard to consume. Radioactive particles in food can enter your body, causing long-term exposure that can lead to serious health risks, including radiation poisoning and cancer. In this chapter, we'll explore how to clean tainted food, understand the risks associated with consuming contaminated items, and learn what steps to take to protect your food supply in a fallout-affected environment.

How Food Becomes Contaminated by Fallout

Radioactive fallout consists of tiny particles of radioactive dust and debris that are released into the atmosphere following a nuclear explosion. These particles can drift for miles, eventually settling on the ground, in water, and on plants and animals. When fallout lands on food, whether it's crops, livestock, or packaged goods, it becomes a direct threat to human health if consumed.

There are two main ways food becomes contaminated:

External Contamination: Fallout particles settle on the surface of food, packaging, or crops. These particles can be removed with proper cleaning methods, but failure to do so can lead to ingestion of radioactive material.

Internal Contamination: This occurs when radioactive particles are absorbed by plants or animals. Crops growing in contaminated soil may absorb radioactive isotopes through their roots, and animals that consume contaminated plants or water can pass these radioactive particles on to humans through their meat or milk. This type of contamination is much more difficult to deal with because it's inside the food itself.

Risks of Consuming Contaminated Food

The ingestion of radioactive particles can be extremely hazardous to your health. Unlike external radiation exposure, where shielding and distance can reduce your risk, ingestion brings radioactive particles directly into your body. The most common radioactive isotopes found in fallout include:

Cesium-137: This isotope can contaminate crops and livestock, and once ingested, it accumulates in muscles, leading to long-term radiation exposure.

Strontium-90: Absorbed by plants from contaminated soil, this isotope mimics calcium in the body and can be deposited in bones, where it continues to emit radiation.

Iodine-131: Often found in milk from contaminated livestock, iodine-131 is absorbed by the thyroid gland, increasing the risk of thyroid cancer.

Ingesting food contaminated with any of these isotopes increases the risk of internal radiation exposure, which can lead to a variety of health issues, including:

Radiation poisoning

Increased risk of cancers, particularly thyroid and bone cancers

Organ damage from long-term exposure

Cleaning Externally Contaminated Food

For food that has been exposed to fallout but has not absorbed radioactive particles internally, cleaning the outer surface can significantly reduce the risk of contamination. Here are the steps to clean tainted food safely:

Handling and Precautions

Before cleaning food, take precautions to avoid spreading contamination:

Wear gloves and a mask: Fallout particles can be dangerous if inhaled or transferred to your skin. Always wear gloves and a mask when handling food that may be contaminated by fallout.

Avoid shaking or disturbing contaminated items: Fallout particles can become airborne if disturbed, increasing the risk of inhalation. Handle food gently to avoid spreading radioactive dust.

Cleaning Fresh Produce

Fruits, vegetables, and other produce that have been exposed to fallout can be cleaned with careful washing. Follow these steps:

Rinse thoroughly with water: Use clean, filtered water to rinse fresh produce. Running water helps to remove fallout particles from the surface. If possible, use a Geiger counter to test the water you're using for contamination before washing.

Peel the skin: For many fruits and vegetables, removing the skin can help eliminate a significant portion of the contamination. Discard the peel in a sealed container far away from your living and food storage areas.

Soak in a baking soda solution: To further reduce contamination, soak the produce in a solution of baking soda and water for 5-10 minutes. Baking soda helps neutralize some radioactive particles and allows for easier removal. Rinse again with clean water after soaking.

Avoid bruised or damaged produce: Fallout particles are more likely to stick to damaged or bruised areas, and these spots can also allow contaminants to penetrate the food more deeply. It's best to discard any produce that appears damaged or overly dirty.

Cleaning Packaged Food

Packaged and canned goods are generally safer than fresh produce since the packaging helps shield the food from fallout particles. However, the outer surfaces of these items can still be contaminated.

Wipe down packaging: Use a clean, damp cloth or a solution of water and mild soap to wipe down the outer packaging of canned goods, bottles, and other packaged food. Be thorough, especially around seams and openings, to remove any particles that may have settled there.

Discard contaminated packaging: If the outer packaging is heavily contaminated or damaged, discard it in a sealed bag or container. Transfer the food into clean containers before storing it.

Milk and Dairy Products

Milk from cows, goats, or other animals that have been exposed to fallout may contain radioactive iodine-131, which can accumulate in the thyroid gland when consumed. Unfortunately, washing or boiling milk does not remove internal contamination.

If you suspect milk is contaminated, it's best to avoid consuming it until it can be tested with a radiation detector. Long-term solutions include feeding livestock uncontaminated food and water or sourcing dairy products from safe areas.

Meat, Fish, and Livestock

Livestock and fish are also at risk of contamination if they consume radioactive fallout through food or water. In cases where the contamination is external, cleaning the surface of the meat may help, but internal contamination is much more problematic.

Trim excess fat: Radioactive particles can accumulate in the fat of animals. When preparing meat, trimming away the fat can help reduce the level of contamination.

Rinse with clean water: If possible, rinse the meat under clean, filtered water before cooking. This can help remove some surface contamination.

Cook thoroughly: Cooking meat does not eliminate radioactive contamination, but it's still essential to ensure that food is cooked thoroughly to reduce other potential health risks, such as bacterial infections.

Preventing Food Contamination

The best way to deal with contaminated food is to prevent it from becoming contaminated in the first place. Here are some strategies to protect your food supply:

Store Food Indoors

Keep your food stored indoors, away from areas where fallout may settle. Sealed containers, cans, and bottles provide the best protection against contamination. Store these items in a basement or interior room if possible, as these areas are less likely to be exposed to fallout.

Grow Food in Greenhouses or Indoor Gardens

For long-term survival, growing your own food may become necessary. In a fallout scenario, outdoor gardening may not be safe due to contamination of the soil and air. Instead, consider growing food in greenhouses or indoor gardens where you can control the environment and reduce exposure to fallout.

Hydroponic systems, which allow you to grow plants in water rather than soil, can also be an effective way to avoid contaminated soil.

Use Clean Water Sources

Water used for drinking and food preparation must be filtered and tested for radioactive contamination. Use a high-quality water filtration system designed to remove radioactive particles, and always test your water with a Geiger counter before using it for cleaning food or cooking.

Testing Food for Radiation

Even after cleaning food, it's important to test it for radiation to ensure it's safe to eat. Using a Geiger counter, you can measure the radiation levels on food surfaces and packaging. If radiation levels are detectable, it may not be safe to consume the food, even after cleaning.

Here's how to test food with a Geiger counter:

Set the Geiger counter to the appropriate sensitivity: Most Geiger counters can detect beta and gamma radiation, which are the most likely forms of radiation present in fallout-contaminated food.

Test food items individually: Place the Geiger counter near the surface of the food or packaging and monitor the reading for several seconds. If radiation levels are higher than normal background levels (typically between 0.05 to 0.20 μSv/h), the food may still be contaminated.

Compare to safe levels: As a rule of thumb, radiation levels in food should not exceed 0.2 μSv/h. If the readings are higher, it's safest to avoid consuming the food, especially if the source of contamination is internal.

In the wake of a nuclear event, cleaning tainted food is a crucial step to ensuring your survival and health. While it is possible to reduce external contamination on many foods by thoroughly cleaning and peeling, internal contamination is much harder to address. By understanding how food becomes contaminated, using proper cleaning techniques, and testing food for radiation, you can significantly reduce the risks associated with fallout exposure. Always prioritize your safety, and if in doubt, avoid consuming food that you suspect may be contaminated.

Purifying Water in Fallout Zones

In the aftermath of a nuclear event, ensuring access to clean, drinkable water is one of the highest priorities for survival. Water sources can easily become contaminated by radioactive fallout, making them unsafe for consumption. Fallout particles can settle on lakes, rivers, reservoirs, and even groundwater, posing a serious risk to human health. Drinking or using contaminated water for cooking or hygiene can lead to internal radiation exposure, increasing the risk of radiation sickness, cancer, and other long-term health issues. In this chapter, we will discuss the methods for purifying water in fallout zones, how to assess water safety, and how to ensure a sustainable supply of clean water in a contaminated environment.

How Water Becomes Contaminated by Fallout

Radioactive fallout consists of tiny particles of dust and debris that are ejected into the atmosphere during a nuclear explosion. These particles eventually settle back to the ground and can land on open bodies of water or seep into groundwater sources. Fallout can also be washed into water systems by rain, further spreading contamination.

There are two primary ways water becomes contaminated by fallout:

External Contamination: Fallout particles settle on the surface of water sources, such as lakes, rivers, and reservoirs. While these particles may float or sink over time, they can still make the water hazardous to drink or use.

Internal Contamination: Over time, radioactive particles can dissolve into the water, making them even harder to remove. Groundwater, wells, and deep aquifers can also become contaminated through radioactive particles seeping into the soil and entering water supplies.

Risks of Drinking Contaminated Water

Drinking water contaminated with radioactive particles poses a significant health risk because it introduces radiation directly into the body. Once inside, radioactive particles can lodge in organs, tissues, and bones, emitting radiation and causing long-term damage. The most common radioactive isotopes found in fallout-contaminated water include:

Iodine-131: This isotope, when ingested, accumulates in the thyroid gland and can cause thyroid cancer or other thyroid-related health issues.

Cesium-137: Cesium tends to accumulate in soft tissues, especially muscles, leading to long-term radiation exposure and an increased risk of cancer.

Strontium-90: Strontium mimics calcium and is absorbed into bones, where it can cause bone cancer and other serious health problems.

Ingesting even small amounts of these radioactive isotopes can lead to dangerous health effects, particularly over time. That's why purifying water in a fallout zone is essential to survival.

Testing Water for Contamination

Before attempting to purify water, it's important to assess its safety by testing for radioactive contamination. A Geiger counter or a specialized water radiation testing kit can help determine if a water source is contaminated with fallout.

Steps for Testing Water

Set the Geiger counter or testing kit to the appropriate sensitivity for detecting beta and gamma radiation, which are common in contaminated water sources.

Collect a water sample in a clean, uncontaminated container. Ideally, take the sample from a depth below the surface, as surface water may have a higher concentration of fallout particles.

Test the sample by holding the Geiger counter near the water or using the radiation testing kit according to the manufacturer's instructions. If the water registers higher-than-normal background radiation levels (typically 0.05 to 0.20 μSv/h), the water may be contaminated.

Assess the results: If the water is contaminated, it will need to be purified or avoided altogether. If no contamination is detected, continue testing periodically, as fallout can settle over time or be introduced through rainfall.

Methods for Purifying Water in Fallout Zones

If water sources are contaminated by fallout, it's critical to purify the water before drinking or using it. Several methods can help remove radioactive particles and make water safe for consumption. However, it's important to note that no single method is perfect, and combining multiple purification techniques may be necessary to ensure water safety.

Sedimentation

Sedimentation is the process of allowing radioactive particles to settle at the bottom of a container, which can then be separated from the clean water at the top. This method works best for removing larger fallout particles from the water but does not eliminate dissolved radioactive materials.

Steps for sedimentation:

Collect water in a large container and allow it to sit undisturbed for 24 to 48 hours.

After the particles settle at the bottom, carefully pour off the clear water from the top, avoiding disturbing the sediment.

Follow up with other purification methods to ensure that any dissolved particles are removed.

Filtration

Filtration is one of the most effective methods for removing fallout particles from water. A high-quality water filter designed to remove radioactive contaminants is essential in fallout zones. There are several types of filters you can use, depending on your situation:

Ceramic filters: These filters have a fine pore structure that can trap fallout particles. While ceramic filters are effective for removing larger particles, they may not filter out dissolved radioactive isotopes.

Activated carbon filters: Carbon filters are excellent for removing dissolved contaminants and certain radioactive particles. They work by trapping particles as water passes through the carbon, which can also reduce the taste and odor of contaminated water.

Reverse osmosis: Reverse osmosis (RO) systems force water through a semi-permeable membrane, which removes a wide range of contaminants, including some radioactive particles. RO systems are highly effective but may require electricity or significant pressure to operate, which could be challenging in a survival scenario.

Boiling

Boiling water is a widely recommended method for killing bacteria and viruses, but it is not effective for removing radioactive particles. In fact, boiling can sometimes concentrate radioactive materials by evaporating water and leaving contaminants behind. Therefore, boiling should not be relied on as a primary method for purifying fallout-contaminated water.

Distillation

Distillation is one of the best methods for removing radioactive particles from water. It works by heating water until it turns into steam, which then condenses into clean, purified water. Since most radioactive particles do not evaporate, they are left behind in the heating container, and the condensed water is much safer to drink.

Steps for distillation:

Heat contaminated water in a pot or distillation setup until it boils.

Capture the steam and direct it into a clean container where it can condense into liquid.

The resulting distilled water should be free from radioactive particles, although it may still lack certain minerals.

Distillation can be a slow process and requires a heat source, but it is highly effective for long-term water purification.

Ion Exchange Resins

Ion exchange resins are specialized materials that can remove specific radioactive isotopes from water. These resins work by exchanging harmful ions (such as cesium or strontium) with harmless ions like sodium or potassium, effectively filtering out radioactive particles.

Applications: Ion exchange systems are often used in large-scale water purification systems, but portable ion exchange filters are also available for individual use. Combining this method with filtration can enhance overall water safety.

Using Bleach or Iodine

While bleach and iodine are effective at disinfecting water by killing bacteria and viruses, they are not capable of removing radioactive contamination. These chemicals can still be used to sanitize water after fallout particles have been filtered out, but they do not purify the water from radiation itself.

Long-Term Water Purification Solutions

In a prolonged fallout scenario, securing a reliable and safe source of water will be critical to survival. Here are some long-term strategies for ensuring access to clean water:

Rainwater Harvesting

Rainwater can be a valuable source of clean water if collected properly. Fallout can contaminate rainwater as it falls through the atmosphere, but after the initial fallout period (about 2 weeks), subsequent rainfalls are likely to be less contaminated.

Steps for Harvesting Rainwater:

Collect rainwater using a clean tarp or other non-contaminated surface, ensuring it's not exposed to fallout particles from nearby structures or vegetation.

Filter and test the collected rainwater for any remaining contamination.

Store rainwater in sealed containers to prevent contamination from future fallout.

Deep Wells and Groundwater

Deep wells that draw water from underground aquifers are less likely to be contaminated by fallout, especially if the aquifer is deep and protected by layers of rock and soil. However, fallout particles can still enter groundwater through cracks in the earth or contaminated rainwater.

Steps for securing a well:

Test the well water regularly with a Geiger counter or water testing kit to ensure it remains uncontaminated.

Use a filtration or purification system to treat well water before drinking, especially if contamination is detected.

Emergency Water Storage

In preparation for a fallout event, storing clean water in advance is one of the best ways to ensure access to safe drinking water. Store water in sealed, food-grade containers in a cool, dark place, away from areas that could be exposed to fallout.

Purifying water in fallout zones is an essential survival skill, as contaminated water poses a significant health risk. By using a combination of sedimentation, filtration, distillation, and other purification methods, you can reduce the presence of radioactive particles and ensure access to clean water. Regularly testing water sources for radiation and taking steps to protect long-term water supplies will help you and your family survive in a fallout-affected environment.

Soil Contamination: The Hidden Dangers of Fallout

In the aftermath of a nuclear event, one of the most insidious and long-lasting effects of fallout is soil contamination. Fallout doesn't just linger in the air; it settles on the ground, becoming embedded in the soil and posing serious, often invisible risks to both human health and the environment. Radioactive particles can remain in the soil for decades, contaminating crops, water supplies, and livestock, and making land uninhabitable for years. In this chapter, we'll explore how soil becomes contaminated by fallout, the dangers it poses, how to assess soil contamination, and what steps you can take to mitigate the risks and reclaim contaminated land for safe use.

How Soil Becomes Contaminated by Fallout

When a nuclear explosion occurs, radioactive particles are blasted into the atmosphere, where they are carried by the wind and eventually settle back to Earth. This process, known as fallout, can contaminate vast areas, far beyond the immediate blast zone. Once radioactive particles settle on the ground, they begin to infiltrate the soil, where they can remain for years, sometimes even centuries.

There are two primary ways that radioactive fallout contaminates soil:

Direct Fallout Deposition: Fallout particles land directly on the soil, where they either remain on the surface or are gradually mixed into deeper layers through natural processes like rain, wind, and the movement of animals or machinery.

Indirect Contamination: Fallout can also enter the soil through contaminated water sources, such as rainfall that has picked up radioactive particles in the atmosphere or surface water that has become polluted by fallout. This water then percolates through the soil, carrying radioactive particles with it.

Once fallout particles are in the soil, they can be absorbed by plants, animals, and humans through the food chain, leading to internal radiation exposure. This type of contamination is particularly dangerous because it is often difficult to detect without proper testing, and it can persist long after the initial fallout event.

Types of Radioactive Contaminants in Soil

Not all radioactive particles behave the same way in soil. Different isotopes have different half-lives and affect soil in unique ways. The most common radioactive contaminants found in soil after a nuclear event include:

Cesium-137: Cesium-137 is one of the most common radioactive isotopes found in nuclear fallout. It has a half-life of about 30 years and binds to soil particles, where it can be absorbed by plants. Cesium-137 tends to accumulate in the muscles of animals and humans, making it a significant long-term health risk if ingested through contaminated food or water.

Strontium-90: Like cesium, strontium-90 is another common isotope in fallout. It mimics calcium and is easily absorbed by plants, which in turn transfer it to the bones of animals and humans. With a half-life of approximately 29 years, strontium-90 poses a serious risk for bone cancer and leukemia if consumed through contaminated crops or animal products.

Plutonium-239: Plutonium is a heavy, highly toxic element with a half-life of 24,100 years. While it tends to stay in the upper layers of soil, it can be inhaled if disturbed, and even small amounts can cause severe damage to the lungs, liver, and bones.

Iodine-131: Though iodine-131 has a much shorter half-life (8 days), it is particularly dangerous in the immediate aftermath of a nuclear event. It is quickly absorbed by the thyroid gland and can lead to thyroid cancer. While it decays relatively quickly, its immediate impact on soil and plants can lead to short-term contamination of food supplies.

The Hidden Dangers of Contaminated Soil

Soil contamination may not be immediately visible, but its dangers are far-reaching. Even in areas where radiation levels seem low in the air, the soil may still be heavily contaminated. The consequences of soil contamination can include:

Contaminated Food Supply

One of the most significant risks of soil contamination is its impact on the food supply. Radioactive isotopes like cesium-137 and strontium-90 can be absorbed by plants through their roots. This contamination can make its way into fruits, vegetables, and grains, which are then consumed by humans or livestock. Even if the surface of the food appears clean, the radioactive particles may be present inside the food itself.

Ingesting contaminated food leads to internal radiation exposure, which can have devastating long-term effects, including cancer, genetic mutations, and other health problems.

Livestock Contamination

Livestock that graze on contaminated soil or drink from polluted water sources are at high risk of absorbing radioactive particles. Dairy products, such as milk from cows or goats, are particularly vulnerable to contamination by iodine-131, which can accumulate in the animals' thyroid glands. Consuming meat, milk, or other products from contaminated animals can expose humans to significant levels of radiation.

Water Contamination

Soil contamination doesn't stay isolated to the ground. Radioactive particles in the soil can leach into groundwater supplies or be carried into surface water sources through runoff during rain or irrigation. This can lead to widespread contamination of water supplies, further compounding the dangers of a nuclear fallout event.

Environmental and Ecological Damage

Soil contamination doesn't just affect humans and livestock—it also has a profound impact on the environment. Radioactive isotopes in the soil can disrupt ecosystems, leading to the death of plants, animals, and insects. Over time, this damage can have a cascading effect, impacting biodiversity, soil fertility, and the ability of land to support life.

Assessing Soil Contamination

Knowing whether the soil in your area is contaminated is critical to determining how to move forward after a nuclear event. Testing the soil for radiation levels is the best way to assess whether it is safe to grow food, raise livestock, or use the land for other purposes.

Using a Geiger Counter

A Geiger counter can give you a general sense of radiation levels in the soil. To use it effectively:

Set the counter to measure beta and gamma radiation, as these are the most common forms of radiation found in contaminated soil.

Test multiple areas: Fallout doesn't settle evenly, so radiation levels can vary significantly within a small area. Test different locations, especially low-lying areas where fallout particles are more likely to accumulate.

Measure at ground level: Hold the Geiger counter close to the soil to get an accurate reading of contamination levels.

While a Geiger counter can detect radiation on the surface of the soil, it may not detect contamination in deeper layers. For more precise testing, professional soil analysis may be required.

Soil Sample Testing

For more accurate results, collect soil samples from different areas of your property and send them to a laboratory for testing. Soil sample testing can identify specific isotopes present in the soil and provide a more detailed understanding of contamination levels.

Collect samples from a variety of depths, as radioactive particles can penetrate deeper into the soil over time.

Test different soil types: Sandy soil, clay, and loam may hold onto radioactive particles differently, so it's important to test a range of soil types in your area.

Mitigating Soil Contamination

While the risks of soil contamination are serious, there are steps you can take to reduce the dangers and reclaim the land for safe use. The process of decontaminating soil may take time, but it can make a significant difference in reducing long-term radiation exposure.

Topsoil Removal

In some cases, the best way to deal with contaminated soil is to remove the top layer, where the fallout particles are most concentrated. This method is especially useful for small, high-priority areas like gardens or crop fields.

Remove 2-4 inches of topsoil: Use a shovel or heavy equipment to carefully remove the top layer of soil, being mindful not to disturb deeper layers unnecessarily.

Dispose of contaminated soil safely: Place the contaminated soil in sealed containers or bags and bury it far from your living areas, ideally in a low-traffic zone where it won't be disturbed.

Soil Dilution

In areas where removing the topsoil isn't practical, soil dilution can help reduce radiation levels. This process involves mixing contaminated soil with clean soil to reduce the concentration of radioactive particles.

Add clean soil: Bring in clean soil from a non-contaminated area and mix it with the contaminated soil to reduce overall radiation levels.

Use cover crops: Planting deep-rooted cover crops, such as clover or alfalfa, can help draw radioactive particles out of the soil. After a season, these crops should be harvested and disposed of as contaminated material.

Phytoremediation (Using Plants to Clean the Soil)

Certain plants have the ability to absorb radioactive particles from the soil through their roots. This process, known as phytoremediation, can be an effective way to gradually clean contaminated soil over time.

Sunflowers, mustard greens, and certain ferns are known for their ability to absorb radioactive isotopes, particularly cesium-137 and strontium-90.

Harvest and dispose of the plants carefully after they've grown to maturity. These plants will contain radioactive particles and should be handled as contaminated material.

Soil Stabilization

If you are unable to decontaminate the soil completely, stabilizing the soil to prevent dust and erosion can help reduce the spread of contamination. Covering contaminated soil with mulch, gravel, or plastic sheeting can prevent radioactive dust from becoming airborne or washing into nearby water sources.

Long-Term Soil Management

Even after decontaminating soil, ongoing management and testing are essential to ensure that radiation levels remain safe. Fallout particles can remain in the soil for years, so regular monitoring will help you track the effectiveness of your decontamination efforts and ensure that the land remains safe for use.

Test the soil annually to monitor radiation levels and detect any new contamination.

Avoid deep plowing or disturbing the soil unnecessarily, as this can bring buried fallout particles back to the surface.

Soil contamination is one of the hidden dangers of nuclear fallout, posing long-term risks to the food supply, the environment, and human health. Understanding how radioactive particles infiltrate soil and taking steps to assess and mitigate contamination can help protect your land and make it safer for future use. By regularly testing soil, removing or diluting contaminated layers, and using phytoremediation techniques, you can reduce the impact of fallout and create a safer environment for yourself and your family in a post-nuclear world.

Growing Food in a Polluted World: Can it be Done?

In a post-nuclear world where radioactive fallout contaminates the environment, the question of whether it's possible to grow safe, nutritious food becomes critical to survival. Fallout affects the air, water, and, most importantly, the soil, which is the foundation of agricultural productivity. The contamination of crops can lead to dangerous levels of internal radiation exposure for anyone consuming them. However, with the right techniques and careful planning, growing food in a polluted world is possible. This chapter explores the challenges of cultivating crops in a fallout-affected environment, strategies for minimizing contamination, and how to adapt traditional gardening methods to ensure food safety.

The Impact of Fallout on Agriculture

Radioactive fallout has a profound effect on agriculture, contaminating both the soil and water that plants rely on to grow. Fallout particles, particularly isotopes like cesium-137, strontium-90, and iodine-131, can be absorbed by plants through their roots and deposited in edible parts such as leaves, fruits, and grains. This contamination poses significant health risks to anyone who consumes these crops, as the radioactive particles continue to emit radiation inside the body long after the food has been eaten.

The challenge in growing food after a nuclear event is finding ways to either avoid contamination or reduce it to safe levels. While complete protection from radiation may not always be possible, there are ways to minimize the risks and still produce a viable food supply.

Assessing Your Growing Environment

Before you begin growing food in a fallout-affected area, it's essential to assess the level of contamination in your environment. This assessment will help you determine whether it's safe to grow food directly in the soil or if alternative growing methods are necessary.

Testing the Soil

As discussed in previous chapters, soil contamination is a major concern in a fallout zone. Using a Geiger counter or sending soil samples to a lab for radiation testing will give you a sense of whether your soil is safe for growing food. Pay particular attention to levels of cesium-137 and strontium-90, as these isotopes are commonly absorbed by plants and pose significant health risks.

If soil contamination is high, it may be necessary to use alternative growing methods, such as raised beds, hydroponics, or indoor gardening, which we'll explore later in this chapter.

Testing Water Sources

Water used for irrigation must also be tested for radioactive contamination. Fallout particles can accumulate in lakes, rivers, and groundwater, making it dangerous to use unfiltered water for crop irrigation. If your water source is contaminated, consider using filtered water or collecting rainwater after the initial fallout period has passed.

Location and Shelter

The location of your garden plays a significant role in minimizing contamination. Growing food in an enclosed space, such as a greenhouse or indoor garden, can help protect plants from fallout particles that settle from the air.

Additionally, choosing a sheltered location—such as a garden near the side of a building or beneath trees—can reduce the amount of fallout that reaches your crops.

Growing Food in Contaminated Soil: Strategies and Techniques

If you determine that growing food in contaminated soil is necessary, there are techniques you can use to minimize the absorption of radioactive particles by your plants.

Raised Beds with Clean Soil

One of the best ways to avoid direct contamination from fallout is by using raised garden beds filled with clean, uncontaminated soil. Raised beds allow you to control the growing medium, reducing the risk of plants absorbing radioactive isotopes from the ground. When building raised beds:

Line the bottom of the bed with a protective barrier, such as plastic sheeting, to prevent contaminated soil from mixing with the clean soil in the bed.

Source soil from uncontaminated areas or purchase soil that has been tested for radiation. This ensures that the growing medium is free from fallout particles.

Raised beds also have the added benefit of providing better drainage, which helps prevent contaminated water from pooling around your crops.

Soil Additives to Reduce Contamination

Certain soil additives can help reduce the absorption of radioactive particles by plants. These substances bind to radioactive isotopes in the soil, making them less available for uptake by plant roots. Some effective soil additives include:

Potassium fertilizers: Potassium competes with cesium-137 for uptake by plant roots. Adding potassium to your soil can reduce the amount of cesium absorbed by crops.

Calcium and phosphorus: Adding calcium to the soil can reduce the uptake of strontium-90, which mimics calcium in biological systems. Phosphorus fertilizers can also help mitigate the absorption of radioactive isotopes.

Compost and organic matter: Increasing the organic matter content in your soil can help dilute radioactive particles, reducing the overall concentration of contamination. Organic matter also improves soil structure and nutrient availability.

Planting the Right Crops

Some plants are more efficient at absorbing radioactive particles from the soil than others. By carefully selecting the types of crops you grow, you can minimize the risk of contamination. For example:

Leafy greens, such as spinach, lettuce, and kale, tend to absorb more radioactive particles than fruiting plants like tomatoes or squash.

Root crops, such as carrots and potatoes, are also prone to absorbing radioactive particles from the soil, particularly cesium-137 and strontium-90.

Fruits and grains, such as corn, wheat, and berries, are generally safer to grow in contaminated soil, as they tend to absorb fewer radioactive particles.

By focusing on crops that are less likely to absorb radiation, you can reduce the overall risk to your food supply.

Alternative Growing Methods: Safe Gardening in Fallout Zones

If soil contamination is too high to safely grow food, or if you want to further reduce the risk of contamination, alternative growing methods offer a viable solution. These methods allow you to grow food in controlled environments, minimizing exposure to fallout.

Greenhouse Gardening

Greenhouses provide an excellent way to protect your crops from airborne fallout particles while maintaining a controlled growing environment. By growing food in a greenhouse, you can shield your plants from contaminated rain, dust, and soil. Additionally, a greenhouse allows you to regulate temperature, humidity, and water quality, which can be crucial in a post-nuclear world where weather patterns and environmental conditions are unpredictable.

Use clean, uncontaminated soil in your greenhouse beds or containers.

Ventilate with filtered air if possible, to prevent fallout particles from entering the greenhouse.

Irrigate with filtered water to ensure your crops remain free of radioactive contamination.

Hydroponics

Hydroponic systems allow you to grow plants without soil, using a nutrient-rich water solution instead. This method eliminates the risk of soil contamination entirely and provides a highly efficient way to produce food in a controlled environment. Hydroponics can be set up indoors or in a greenhouse, further reducing the risk of exposure to fallout particles.

Use filtered water in your hydroponic system to avoid introducing radioactive particles.

Monitor nutrient levels closely to ensure your plants receive the necessary minerals without relying on contaminated soil.

Hydroponic systems can be resource-intensive, requiring electricity for pumps and lights, but they offer a reliable way to grow food in a fallout-affected world.

Indoor Gardening

For those with limited access to outdoor space or clean soil, indoor gardening is a practical solution. Growing food indoors allows you to completely control the environment, protecting your crops from fallout and other contaminants. You can grow a wide variety of fruits, vegetables, and herbs indoors using containers, raised beds, or hydroponic systems.

Use artificial lighting, such as LED grow lights, to provide the necessary light for photosynthesis.

Irrigate with filtered water and use clean, uncontaminated soil or hydroponic nutrients to nourish your plants.

Maintain proper ventilation to ensure your indoor garden remains healthy and free from mold or other issues.

Indoor gardening requires careful planning and resource management, but it can provide a steady supply of food in a highly contaminated environment.

Managing and Testing Your Food Supply

Even with careful planning and protective growing techniques, it's essential to regularly test your food supply for radiation to ensure it's safe to eat. Use a Geiger counter or a specialized food testing kit to monitor radiation levels in your crops before consuming them.

Test harvested produce by holding the Geiger counter close to the surface of the fruit, vegetable, or grain. If radiation levels are higher than normal background levels, it may not be safe to eat.

Focus on long-term food storage: To ensure food safety, consider storing non-perishable items such as canned goods, dried foods, and preserved fruits and vegetables. Stocking up on clean, uncontaminated food before a nuclear event can help supplement your diet and reduce the need to rely solely on home-grown crops.

Growing food in a world polluted by radioactive fallout is challenging, but it's not impossible. By carefully assessing your environment, using raised beds or alternative growing methods, and selecting the right crops, you can minimize contamination and produce safe, nutritious food. Whether you choose to garden in a greenhouse, set up a hydroponic system, or grow crops indoors, the key to success is controlling your growing conditions and testing your food regularly to ensure its safety. With careful planning and the right strategies, you can thrive in even the most difficult conditions and ensure a steady food supply for the long term.

Radiation and Plants: Understanding the Risks

In a post-nuclear world, radiation poses a significant risk to plant life, particularly in areas affected by fallout. Plants absorb nutrients and water from the soil, making them vulnerable to radioactive particles that may have settled there. Understanding how radiation impacts plants, the risks it poses to food production, and the long-term effects on ecosystems is essential for anyone attempting to grow food in a contaminated environment. In this chapter, we will explore how radiation affects plants, the risks associated with growing food in a radioactive zone, and the steps you can take to mitigate those risks.

How Radiation Affects Plants

Radiation can have a variety of effects on plants, depending on the type of radiation, the intensity of exposure, and the duration of exposure. The most common radioactive isotopes found in fallout—cesium-137, strontium-90, and iodine-131—pose specific risks to plants, particularly in how they are absorbed and concentrated within different parts of the plant.

Absorption of Radioactive Particles

Radioactive fallout can settle on plants in two primary ways: by landing directly on the leaves and stems, or by being absorbed through the roots from contaminated soil and water. Both routes can lead to the accumulation of radioactive particles within the plant's tissues.

Surface contamination: Fallout that settles on the surface of plants, such as leaves and fruits, can be washed off with proper cleaning methods (as discussed in previous chapters). However, if not removed, these particles can be ingested when the plant is consumed, posing a risk to human health.

Root absorption: The more serious threat comes from plants absorbing radioactive particles through their roots. Once inside the plant, radioactive isotopes like cesium-137 and strontium-90 mimic essential nutrients (such as potassium and calcium) and become integrated into the plant's tissues. This contamination cannot be removed by washing or peeling and poses a long-term health risk when the plant is consumed.

Effects on Plant Growth and Development

Radiation can disrupt the normal growth and development of plants. High levels of radiation exposure can lead to visible damage, such as stunted growth, leaf discoloration, and reduced crop yields. In severe cases, radiation exposure can cause mutations in plants, leading to abnormal growth patterns or even the death of the plant.

Cellular damage: Radiation can damage plant cells by breaking chemical bonds in the plant's DNA. This can lead to mutations that alter the plant's growth and reproductive capabilities. In some cases, these mutations may be harmful, leading to deformities or a reduction in the plant's ability to produce viable seeds.

Delayed germination: Radiation exposure can slow down or prevent the germination of seeds. Even if the seeds sprout, plants may develop more slowly or produce fewer fruits and vegetables than they would in a radiation-free environment.

Long-term genetic mutations: Just as radiation can cause mutations in animals and humans, it can also lead to genetic changes in plants. These mutations may affect future generations of plants, potentially leading to reduced viability or unpredictable traits.

Accumulation of Radioactive Isotopes

Different parts of a plant can accumulate radioactive particles at varying concentrations. Understanding where radiation accumulates within a plant can help you make informed decisions about which parts are safer to consume.

Roots: Plants with large root systems, such as potatoes, carrots, and other root vegetables, are more likely to absorb radioactive particles from contaminated soil. These crops may accumulate higher concentrations of cesium-137 and strontium-90.

Leaves: Leafy greens like spinach, lettuce, and kale are particularly vulnerable to surface contamination, as fallout particles can easily settle on their broad leaves. However, leafy plants may also absorb radioactive particles through their roots, making the entire plant potentially hazardous.

Fruits and grains: Fruits, grains, and seeds generally absorb less radiation than leaves and roots. However, they can still be contaminated if grown in radioactive soil, particularly if radioactive isotopes are present in the water used for irrigation.

Risks of Growing Food in a Fallout Zone

Growing food in a fallout zone comes with inherent risks due to the potential for radioactive contamination. While it may be possible to grow crops safely with the right precautions, it's important to understand the risks involved and to take steps to minimize exposure to radiation.

Internal Radiation Exposure

When you consume food contaminated with radioactive particles, those particles are absorbed into your body and continue to emit radiation from within. This type of internal radiation exposure is particularly dangerous because it can lead to long-term damage to cells and tissues. The most common radioactive isotopes found in fallout, such as cesium-137 and strontium-90, can accumulate in the body's soft tissues and bones, increasing the risk of cancer and other serious health issues.

Long-Term Contamination

Radioactive particles in the soil and water don't disappear quickly. Some isotopes, like iodine-131, have relatively short half-lives (around 8 days), meaning they decay quickly. Others, such as cesium-137 and strontium-90, have half-lives of around 30 years, meaning they will continue to pose a threat for decades. This long-term contamination makes it essential to monitor soil and water regularly and use protective measures to reduce the risk of contamination in crops.

Food Supply Chain Contamination

Even if you're careful to grow your own food in a controlled environment, contamination can enter your food supply from other sources. For example, livestock that graze on contaminated land or drink from polluted water sources may accumulate radioactive particles in their meat, milk, and other products. Contaminated food can also make its way into markets, posing a risk even to those who are not growing their own crops.

Mitigating the Risks of Growing Food in a Radioactive Environment

While growing food in a fallout zone presents challenges, it is possible to mitigate many of the risks with careful planning and the right strategies. Here are some steps you can take to reduce the impact of radiation on your crops.

Use Raised Beds and Clean Soil

One of the most effective ways to reduce radiation exposure in plants is to grow them in raised beds filled with clean, uncontaminated soil. Raised beds help to isolate your plants from contaminated ground soil and allow you to control the growing environment.

Choose clean, uncontaminated soil from a safe source or purchase soil that has been tested for radiation.

Line the bottom of the raised beds with a protective barrier, such as plastic sheeting, to prevent contamination from seeping in from the ground.

Irrigate with Filtered Water

Water contamination is a serious concern in a fallout zone, as radioactive particles can enter the soil through irrigation. Using filtered water for irrigation can help reduce the amount of radioactive contamination your plants are exposed to.

Use water filtration systems that are capable of removing radioactive particles, such as reverse osmosis or activated carbon filters.

Collect and store rainwater for use in irrigation, but only after the initial fallout period has passed and the risk of contamination from fallout particles in the air has decreased.

Choose Safer Crops

Certain crops are more likely to absorb radioactive particles than others. By choosing crops that are less prone to contamination, you can reduce the risk of growing food in a radioactive environment.

Fruiting plants: Crops that produce fruits, such as tomatoes, peppers, and squash, tend to absorb less radiation than leafy greens or root vegetables.

Grains and legumes: Wheat, barley, beans, and peas are generally less likely to accumulate high levels of radiation compared to root crops like carrots and potatoes.

Use Soil Amendments to Block Absorption

Adding specific soil amendments can help reduce the uptake of radioactive particles by plants. For example, adding potassium to the soil can help block the absorption of cesium-137, while adding calcium can help reduce the absorption of strontium-90.

Apply potassium-rich fertilizers to the soil to reduce cesium uptake by plants.

Add calcium and phosphorus to reduce the absorption of strontium-90.

Regularly Test Your Crops for Radiation

Regular testing of your crops is essential to ensure that the food you are growing is safe to eat. Use a Geiger counter or specialized food testing kit to monitor radiation levels in your plants before consuming them.

Test crops at harvest: Pay special attention to plants that are more prone to radiation absorption, such as leafy greens and root vegetables.

Use multiple testing methods: Testing both the soil and the plants themselves will give you a clearer picture of the overall contamination risk.

Long-Term Effects on Plant Life and Ecosystems

While the immediate effects of radiation on plants can be severe, the long-term impacts on ecosystems are equally concerning. Over time, radioactive particles in the soil can disrupt entire food chains, leading to reduced biodiversity and altered ecological balance. Forests, grasslands, and wetlands may experience long-term degradation due to the death of plants, animals, and microorganisms that are essential to healthy ecosystems.

In areas with heavy contamination, it may take decades or even centuries for the environment to recover fully. However, some plant species may be more resistant to radiation than others, and these species could play a crucial role in the regeneration of ecosystems over time.

Radiation poses significant risks to plants, both in terms of their growth and development and the potential for contamination of the food supply. However, by understanding how radiation affects plants and taking steps to mitigate these risks, it is possible to grow food safely in a fallout-affected world. By using raised beds, filtered water, and careful crop selection, you can reduce radiation exposure and produce food that is safe to eat. Additionally, regular testing of both soil and crops will help ensure that your food supply remains free from harmful levels of contamination.

Water Filtration: Making Contaminated Water Safe

Access to clean, safe water is essential for survival in any environment, but it becomes even more critical in the aftermath of a nuclear event. Radioactive fallout can contaminate water sources, turning otherwise reliable water supplies into a serious health risk. Drinking or using contaminated water can lead to internal radiation exposure, which can cause long-term health issues such as radiation sickness, cancer, and organ damage. In this chapter, we will explore the importance of water filtration in fallout zones, different filtration methods, and how to effectively make contaminated water safe for drinking and everyday use.

The Dangers of Radioactive Contaminants in Water

After a nuclear event, fallout particles can settle on surface water sources such as lakes, rivers, and reservoirs. Additionally, fallout can seep into groundwater supplies, contaminating wells and aquifers. The most dangerous radioactive isotopes commonly found in fallout include:

Cesium-137: A common byproduct of nuclear fission, cesium-137 can dissolve in water and accumulate in the body's soft tissues when ingested, leading to long-term radiation exposure.

Strontium-90: This isotope mimics calcium in the body, and once consumed, it accumulates in bones, increasing the risk of bone cancer and leukemia.

Iodine-131: Though it has a short half-life, iodine-131 is highly dangerous in the immediate aftermath of a nuclear event. It is absorbed by the thyroid gland and can lead to thyroid cancer when consumed in contaminated water or food.

Water contaminated with these radioactive particles is not safe to drink or use for hygiene without proper filtration. Even small amounts of radioactive contamination can lead to serious health risks if consumed regularly, so it's essential to filter and purify any water before using it.

Understanding Water Filtration and Its Importance

Water filtration is the process of removing harmful contaminants, such as radioactive particles, chemicals, and microorganisms, from water to make it safe for drinking and everyday use. Different filtration methods are designed to target specific types of contaminants, so understanding the strengths and limitations of each method is important for selecting the right filtration system for a fallout environment.

While no single filtration method can remove all types of radioactive particles, combining multiple techniques can greatly improve water safety. The goal is to reduce the overall concentration of radioactive contaminants to a level that is considered safe for human consumption.

Methods for Filtering and Purifying Water in Fallout Zones

There are several filtration and purification methods available, each with varying levels of effectiveness in removing radioactive contaminants. Depending on the level of contamination and the resources available, you may need to use a combination of these methods to ensure your water is safe to drink.

Sedimentation and Settling

Sedimentation is the process of allowing heavier radioactive particles to settle at the bottom of a container, leaving cleaner water at the top. While sedimentation alone won't purify water completely, it can help reduce the overall concentration of fallout particles in the water.

Steps for sedimentation:

Collect water in a large, clean container and let it sit undisturbed for 24 to 48 hours.

After the radioactive particles have settled to the bottom, carefully pour off the clear water from the top, leaving the sediment behind.

Follow up with additional filtration and purification methods to further remove dissolved radioactive particles.

Sedimentation is an effective first step in the purification process but should not be relied on as the sole method of filtration, especially in heavily contaminated water sources.

Filtration Systems

Filtration systems are one of the most effective ways to remove radioactive particles and other contaminants from water. Several types of filters can be used in fallout zones, each with its own strengths and limitations:

Activated Carbon Filters: These filters are highly effective at removing dissolved contaminants, such as radioactive iodine-131, and can also reduce chemicals and toxins in the water. Activated carbon works by adsorbing contaminants onto the surface of the carbon particles, trapping them as water passes through.

Strengths: Effective at removing dissolved radioactive particles, chemicals, and organic pollutants.

Limitations: Less effective at removing radioactive particles like cesium-137 and strontium-90. Should be used in conjunction with other filtration methods for best results.

Ceramic Filters: Ceramic filters have a fine pore structure that traps larger particles, including some radioactive fallout particles. These filters are effective for removing sediment, bacteria, and some radioactive materials, though they may not remove dissolved isotopes effectively.

Strengths: Effective for removing larger radioactive particles, bacteria, and sediment.

Limitations: May not remove dissolved radioactive isotopes. Works best when combined with carbon filtration or other purification methods.

Reverse Osmosis (RO) Systems: Reverse osmosis is one of the most effective filtration methods for removing a wide range of contaminants, including radioactive particles. RO systems force water through a semi-permeable membrane, which removes dissolved salts, minerals, and radioactive isotopes like cesium-137 and strontium-90. Reverse osmosis is highly effective but can be resource-intensive, as it requires water pressure and may waste some water in the process.

Strengths: Highly effective at removing dissolved radioactive particles, including cesium-137 and strontium-90.

Limitations: Requires electricity or water pressure to operate, which may be a challenge in survival situations. Produces some wastewater as part of the filtration process.

Distillation

Distillation is a highly effective method for purifying water in fallout zones because it removes nearly all types of radioactive particles, including dissolved isotopes. Distillation works by heating water until it turns into steam, which is then captured and condensed into a separate container, leaving contaminants behind.

Steps for distillation:

Heat contaminated water in a pot or distillation setup until it boils.

Capture the steam and direct it into a clean container where it can condense into liquid.

The resulting distilled water should be free of radioactive particles, though it may lack certain minerals.

Distillation is particularly useful because most radioactive particles do not evaporate with the water, making the resulting water much safer to drink. However, the process can be slow and requires a reliable heat source, which may be difficult in a survival situation.

Ion Exchange Filters

Ion exchange filters are specialized devices that can remove specific radioactive isotopes from water. They work by exchanging harmful ions, such as cesium-137 and strontium-90, for harmless ions like sodium or potassium. Ion exchange resins are often used in large-scale water purification systems, but portable ion exchange filters are also available for individual use.

Strengths: Effective at removing specific radioactive particles like cesium-137 and strontium-90.

Limitations: Not all ion exchange filters are designed for radioactive contamination, so it's important to select one specifically for this purpose. Should be used in combination with other filtration methods for comprehensive water purification.

Boiling

Boiling water is an effective method for killing bacteria, viruses, and other pathogens, but it does not remove radioactive particles from water. In fact, boiling contaminated water may actually concentrate radioactive particles by evaporating water and leaving the contaminants behind. For this reason, boiling should not be relied upon as a purification method in fallout zones unless you are certain the water is free of radioactive contamination.

Building a Comprehensive Water Filtration Strategy

To ensure your water is safe to drink, you may need to use a combination of filtration and purification methods. Here's a step-by-step approach to purifying water in a fallout-affected environment:

Start with Sedimentation

Allow water to sit undisturbed for at least 24 hours to let larger fallout particles settle to the bottom. Carefully pour off the clear water and discard the sediment.

Use a Primary Filtration System

Next, filter the water through a system that is capable of removing radioactive particles. A reverse osmosis system, ceramic filter, or activated carbon filter can be highly effective at this stage.

Use a Secondary Filtration System

For added safety, pass the water through a secondary filter, such as an ion exchange filter or a second carbon filter, to remove any remaining dissolved radioactive particles.

Consider Distillation for Critical Use

If you need the highest level of water purity—such as for drinking water or preparing food—distillation is the most effective method. Use distillation in combination with other filtration techniques to ensure your water is free of both radioactive particles and other contaminants.

Testing Water for Radiation

Even after filtering water, it's important to test it for radiation to ensure it's safe to drink. Use a Geiger counter or a specialized water testing kit to measure radiation levels. Here's how to test water for contamination:

Test the water source: Before collecting water for filtration, test the water at its source (lake, river, well) to determine the level of contamination. If radiation levels are high, consider finding an alternative water source.

Test filtered water: After filtration, test the water again to ensure that radiation levels have dropped to a safe level. If the water still shows elevated radiation, consider running it through another filtration process or using distillation.

Safe radiation levels in water typically range from 0.05 to 0.20 μSv/h (microsieverts per hour). Anything higher may still pose a risk and should be treated further.

Long-Term Water Supply Solutions

For long-term survival in a fallout zone, it's essential to secure a reliable and sustainable water supply. Consider these strategies to ensure access to clean water:

Rainwater Harvesting

Rainwater, especially after the initial fallout period, can be a valuable source of clean water. Collect rainwater using a clean, uncontaminated surface, such as a tarp or roof, and filter it before use. Store rainwater in sealed containers to prevent contamination.

Deep Wells

If you have access to a deep well or aquifer, this may be a safer source of water than surface water. However, test the water regularly, as fallout particles can seep into groundwater supplies over time. Use a filtration system capable of removing radioactive contaminants to ensure the well water remains safe to drink.

In a fallout-affected world, water filtration is critical for survival. By understanding the different filtration methods available and how to use them effectively, you can ensure access to clean, safe drinking water even in the most challenging conditions. Combining multiple filtration techniques—such as sedimentation, activated carbon filters, reverse osmosis, and distillation—will provide the best protection against radioactive contaminants. Regular testing of both water sources and filtered water is essential to ensure your filtration system is working effectively, allowing you to stay safe and healthy in a post-nuclear environment.

Food Storage for the Long Haul

In a world affected by nuclear fallout, securing a reliable and uncontaminated food supply is one of the most important steps for long-term survival. Fallout particles can contaminate crops, livestock, and water, making it essential to carefully plan how you store and protect your food to ensure it remains safe over time. In this chapter, we'll explore strategies for food storage that can sustain you and your family in a fallout zone, focusing on techniques for maximizing shelf life, minimizing contamination risks, and ensuring a balanced diet when access to fresh food may be limited.

Why Long-Term Food Storage is Critical in Fallout Zones

After a nuclear event, access to fresh food can become highly unpredictable or impossible for extended periods. Fallout can contaminate local crops, livestock, and water sources, and it may take months or even years before it is safe to grow fresh produce again in contaminated areas. For this reason, having a long-term food storage plan is critical to survival.

Properly stored food can provide essential nutrition and help you avoid the risks of eating contaminated food. By stockpiling and preserving a variety of food items, you can ensure that you have a reliable food source while avoiding exposure to radioactive particles.

Choosing the Right Foods for Long-Term Storage

Not all foods are suitable for long-term storage, especially in environments where resources like refrigeration or fresh water may be limited. The key to successful long-term food storage is selecting foods that have a long shelf life, require minimal preparation, and provide balanced nutrition. Here are some of the best options:

Dried and Dehydrated Foods

Dried and dehydrated foods are excellent for long-term storage because they are lightweight, have a long shelf life, and retain most of their nutritional value. Removing moisture from food significantly extends its shelf life by preventing the growth of bacteria, mold, and other spoilage organisms.

Dried fruits and vegetables: These provide essential vitamins and fiber. Dehydrated vegetables can be rehydrated and used in soups, stews, and other meals.

Dried beans and legumes: High in protein and fiber, beans and lentils are excellent staples for long-term storage. They can be cooked and combined with rice or other grains for a complete meal.

Powdered eggs and milk: Powdered versions of eggs and milk can be stored for years and rehydrated when needed, providing important protein and calcium.

Instant rice, pasta, and grains: These can be easily rehydrated with water and cooked quickly, making them ideal for emergency meals.

Canned Foods

Canned foods are a critical component of any long-term food storage plan. Canning preserves food by sealing it in airtight containers, preventing contamination and spoilage. Properly stored canned goods can last for years and offer a variety of essential nutrients.

Canned vegetables and fruits: These provide vitamins and fiber, and they can be used in a variety of recipes or eaten straight from the can.

Canned meats and fish: Items like tuna, chicken, and beef provide protein and fats. They are also easy to incorporate into stews, casseroles, or eaten on their own.

Canned soups and stews: These ready-to-eat meals are convenient and can provide a balanced meal with minimal preparation.

Freeze-Dried Foods

Freeze-dried foods are among the best options for long-term storage because they retain most of their nutritional value and flavor while having an extremely long shelf life—often up to 25 years or more. Freeze-drying removes moisture from food through sublimation, preserving it without the need for refrigeration.

Freeze-dried meals: Pre-packaged freeze-dried meals are designed for long-term storage and are convenient to prepare—simply add hot water.

Freeze-dried fruits and vegetables: These make excellent snacks or can be rehydrated and used in cooking.

Freeze-dried proteins: Freeze-dried meats, eggs, and dairy products are lightweight and can be rehydrated for meals that provide essential protein and fats.

Grains and Legumes

Whole grains and legumes are some of the most reliable staples for long-term food storage due to their long shelf life, high nutritional content, and versatility in cooking.

Rice, oats, wheat, and barley: These grains can be stored in bulk and provide carbohydrates, fiber, and essential nutrients. When combined with beans or lentils, they offer complete protein sources.

Dried beans, peas, and lentils: These legumes are rich in protein and fiber and can be stored for years in proper conditions.

Honey, Salt, and Sugar

Honey, salt, and sugar are valuable for long-term food storage because they have an indefinite shelf life when stored properly. Honey is a natural preservative and a source of energy, while salt is essential for preserving other foods and for seasoning. Sugar, in addition to being a sweetener, can be used in baking and food preservation.

Nutritional Supplements

In an environment where access to fresh food is limited, it may be difficult to get all the essential vitamins and minerals from stored food alone. Consider adding multivitamins or specific supplements to your food storage to ensure that you and your family receive balanced nutrition.

Proper Storage Techniques for Long-Term Food Security

Once you've selected the right foods for long-term storage, it's essential to store them properly to maximize shelf life and minimize the risk of contamination. Here are some key storage techniques:

Use Airtight Containers

The most important factor in preventing food spoilage is keeping moisture, air, and pests away from your stored food. Using airtight containers, such as food-grade buckets, mylar bags, or vacuum-sealed jars, helps protect your food from the elements.

Mylar bags: These durable, reflective bags are ideal for storing dried foods. They provide an airtight seal when used with oxygen absorbers and can block out light, which helps preserve food quality.

Vacuum-sealed jars: These are ideal for smaller quantities of dried food and can be stored in a cool, dark place to extend shelf life.

Store in Cool, Dark Places

Temperature and light play a significant role in food preservation. Most stored food should be kept in a cool, dark, and dry environment, such as a basement, cellar, or storage room. The ideal storage temperature is between 50°F and 70°F (10°C to 21°C). Avoid areas where temperature fluctuates widely, as this can accelerate spoilage.

Avoid moisture: High humidity can lead to mold growth and spoilage. Use desiccants in storage containers to absorb any excess moisture.

Rotate Your Stock

To avoid wasting food due to spoilage or expiration, it's important to use the "first in, first out" method for food storage. This means using the oldest items in your stock first and replacing them with new ones. By rotating your stock regularly, you can ensure that your food remains fresh and usable over the long term.

Label and Date All Items

Clearly labeling and dating all stored food is essential for effective stock management. This allows you to keep track of expiration dates and know when to rotate or replace items. Include information about the contents, packaging date, and recommended use-by date on each container.

Use Oxygen Absorbers and Desiccants

Oxygen and moisture are the two main enemies of long-term food storage. To prevent spoilage, place oxygen absorbers or desiccants in your food storage containers. Oxygen absorbers remove oxygen from sealed containers, which helps prevent the growth of bacteria, mold, and other spoilage organisms. Desiccants help absorb moisture, keeping your food dry and safe.

Protect Against Pests

Pests such as rodents, insects, and bacteria can compromise your stored food supply. Use heavy-duty, sealed containers to prevent pests from accessing your food. Additionally, keeping food storage areas clean and free of spills or crumbs will help minimize the risk of attracting pests.

Alternative Food Preservation Methods

In addition to storing canned and dried foods, you can extend your food supply by preserving fresh food using traditional preservation methods. These techniques allow you to make the most of fresh food during growing seasons or before food becomes contaminated by fallout.

Canning

Canning is a tried-and-true method for preserving fruits, vegetables, and meats. By sealing food in airtight jars and heating them to kill bacteria, you can store canned goods for years. Pressure canning is particularly effective for low-acid foods like meats and vegetables.

Dehydrating

Dehydrating fruits, vegetables, and meats removes moisture and significantly extends their shelf life. You can use a commercial dehydrator or dry food in the sun or oven. Once dried, store the food in airtight containers for long-term use.

Pickling

Pickling is a method of preserving food in vinegar or a brine solution. This process prevents bacterial growth and allows certain foods to be stored for extended periods. Pickled vegetables, fruits, and even meats can add variety to your food storage.

Fermentation

Fermentation is another preservation method that uses naturally occurring bacteria to extend the shelf life of foods. Fermented foods, such as sauerkraut, kimchi, and yogurt, can be stored for months and provide beneficial probiotics that promote digestive health.

Protecting Your Food from Fallout Contamination

In a fallout zone, protecting your stored food from radioactive contamination is essential. Fallout particles can settle on surfaces and contaminate food through contact or exposure to air and dust. To prevent contamination:

Store food in sealed containers: Airtight containers will keep fallout particles from entering your food supply. Plastic buckets, mylar bags, and vacuum-sealed jars are excellent options for protecting food from contamination.

Keep food storage areas clean and isolated: Set aside a dedicated area for food storage, away from fallout exposure. Clean surfaces regularly to prevent contamination from dust or airborne fallout particles.

Consider underground storage: If possible, storing food underground in a bunker or basement can provide extra protection from fallout by shielding your stockpile from radiation.

Planning for a Balanced Diet in Survival Situations

In a long-term survival scenario, it's important to ensure that your stored food provides balanced nutrition. Relying solely on one type of food, such as rice or canned goods, can lead to nutritional deficiencies over time. To maintain health and energy, your stored food should include:

Carbohydrates: Grains like rice, oats, and pasta provide the energy you need for daily activities.

Proteins: Canned meats, beans, legumes, and powdered proteins are essential for muscle maintenance and overall health.

Fats: Fats are necessary for energy and cell function. Include items like nuts, seeds, and freeze-dried dairy products in your storage.

Vitamins and minerals: Dried fruits, vegetables, and multivitamins can help provide the vitamins and minerals you may lack when access to fresh food is limited.

Long-term food storage is a crucial component of survival in a fallout-affected world. By selecting the right foods, using proper storage techniques, and protecting your food from contamination, you can ensure that you and your family have a reliable, nutritious food supply for the long haul. Combining canned goods, freeze-dried foods, grains, and dehydrated items with alternative preservation methods like pickling, canning, and fermentation will allow you to maintain a diverse and balanced diet, even in challenging conditions. Regularly testing and rotating your stock will help ensure that your stored food remains safe and usable over time, giving you peace of mind in uncertain times.

Radiation Poisoning: Symptoms and First Aid

Radiation poisoning, also known as acute radiation syndrome (ARS), occurs when a person is exposed to a high dose of ionizing radiation over a short period of time. This can happen during nuclear explosions, reactor accidents, or from exposure to fallout in a post-nuclear environment. The severity of radiation poisoning depends on the dose, duration, and type of radiation exposure, and symptoms can range from mild to life-threatening. In this chapter, we will explore the symptoms of radiation poisoning, the different stages of ARS, and the essential first aid steps that can help manage the condition. While professional medical treatment is necessary in severe cases, knowing how to administer basic first aid can make a critical difference in a survival situation.

What Causes Radiation Poisoning?

Radiation poisoning occurs when ionizing radiation—such as gamma rays, X-rays, or particles from radioactive isotopes—damages or destroys cells in the human body. This damage can affect the body's ability to repair itself, leading to widespread cellular injury, organ failure, and, in extreme cases, death. Radiation primarily harms rapidly dividing cells, such as those in the skin, bone marrow, gastrointestinal tract, and reproductive organs, which is why these areas are often the first to show symptoms.

The sources of radiation exposure that can lead to ARS include:

Direct radiation from a nuclear explosion: Gamma rays and neutrons emitted during the detonation of a nuclear bomb can cause acute radiation poisoning.

Radioactive fallout: Fallout from a nuclear explosion or reactor accident can settle on the ground, water, and objects, emitting radiation that can be absorbed by the body over time.

Contaminated food and water: Consuming radioactive particles through food and water can lead to internal radiation exposure, which can be equally dangerous as external exposure.

Radiation Dose and Its Effects

The effects of radiation exposure depend on the dose received, measured in sieverts (Sv) or millisieverts (mSv). Understanding the relationship between radiation dose and its impact on the body is critical for assessing the severity of radiation poisoning.

0.05 to 0.2 mSv: Typical background radiation from the environment; this level is generally safe and not harmful.

0.2 to 1 Sv: Mild radiation sickness is possible with prolonged exposure, but severe health effects are unlikely.

1 to 2 Sv: Symptoms of radiation sickness begin to appear within hours to days of exposure. Nausea, vomiting, and fatigue are common.

2 to 6 Sv: Severe radiation poisoning occurs at this level, with a high risk of death if untreated. Symptoms include vomiting, diarrhea, and hair loss, with damage to the bone marrow and gastrointestinal tract.

6 Sv and above: Lethal radiation doses. Exposure at this level is often fatal within days to weeks.

Symptoms progress rapidly, including organ failure, severe internal bleeding, and neurological damage.

Symptoms of Radiation Poisoning

Radiation poisoning progresses in stages, with symptoms varying depending on the level of exposure. Identifying these symptoms early is critical for administering timely first aid and seeking medical help.

Stage 1: Prodromal Stage (Initial Symptoms)

The prodromal stage occurs within minutes to hours after exposure and can last for several days. Symptoms are often nonspecific, making it difficult to immediately determine that radiation poisoning is the cause.

Nausea and vomiting: One of the earliest signs of radiation poisoning. The onset of vomiting can be an indicator of the severity of exposure—the sooner vomiting begins after exposure, the higher the dose of radiation.

Fatigue and weakness: Radiation damages cells in the body, leading to extreme tiredness and lack of energy.

Headache: Radiation exposure can cause headaches and dizziness due to the impact on the nervous system.

Loss of appetite: A person exposed to high levels of radiation may lose their appetite as the gastrointestinal system becomes affected.

Stage 2: Latent Stage (Apparent Recovery)

During the latent stage, symptoms may appear to subside, leading the person to believe they are recovering. However, this stage is deceptive, as internal damage to cells and organs continues to progress. The latent stage can last anywhere from a few hours to several weeks, depending on the radiation dose.

Temporary relief of symptoms: After the initial bout of nausea, vomiting, and fatigue, the person may feel better for a short period. However, this is usually followed by a more severe stage of illness.

Stage 3: Manifest Illness Stage (Critical Symptoms)

In the manifest illness stage, symptoms return with increased severity as the body's immune system and organs begin to fail. This stage can last for weeks and is marked by specific symptoms depending on the dose and which systems in the body are affected.

Gastrointestinal symptoms: Severe nausea, vomiting, diarrhea, and abdominal pain occur due to the destruction of cells in the digestive tract. Dehydration and electrolyte imbalances can follow, worsening the condition.

Infections: Radiation damages the immune system, particularly the bone marrow, which produces white blood cells. This leaves the person highly susceptible to infections, which can quickly become life-threatening.

Internal bleeding: Damage to the bone marrow also affects red blood cell and platelet production, leading to anemia and the inability to clot blood. This can result in internal bleeding, including nosebleeds, bleeding gums, and bruising.

Hair loss: Radiation targets rapidly dividing cells, including hair follicles, causing significant hair loss.

Skin burns and lesions: Skin exposed to high levels of radiation can develop burns, blisters, and lesions over time.

Neurological symptoms: At extremely high doses, radiation can affect the brain, leading to confusion, dizziness, tremors, and seizures.

Stage 4: Recovery or Death

For those exposed to lower doses of radiation, recovery may be possible with proper medical treatment. The body can slowly regenerate some of the damaged cells over weeks to months. However, individuals exposed to higher doses (above 6 Sv) often experience rapid deterioration and death due to multi-organ failure, severe infections, or internal bleeding.

First Aid for Radiation Poisoning

In a survival situation, access to professional medical care may be limited or delayed. Knowing how to provide first aid for radiation poisoning can help manage symptoms and prevent complications. Here are the essential steps for first aid in cases of radiation exposure:

Remove the Person from the Source of Radiation

The first step in providing first aid for radiation poisoning is to remove the person from the source of radiation as quickly as possible. Distance, shielding, and time are the key factors in reducing radiation exposure:

Distance: Move as far away from the radiation source as possible. Even a small increase in distance can significantly reduce exposure.

Shielding: Find shelter in a location that offers protection from radiation, such as a basement, concrete building, or underground bunker. Thick walls, earth, and dense materials like lead can block or reduce radiation levels.

Time: Limit the amount of time spent near the radiation source to minimize exposure.

Decontaminate the Person

If the person has been exposed to radioactive fallout, decontamination is critical to prevent further radiation absorption. This involves removing contaminated clothing and washing the skin to remove fallout particles.

Remove clothing: Carefully remove the person's clothing, which may contain up to 90% of the radioactive particles. Place the clothing in a sealed plastic bag and store it away from living areas.

Wash the skin: Gently wash the exposed skin with soap and lukewarm water to remove fallout particles. Avoid scrubbing too hard, as this could damage the skin and allow radioactive particles to enter the bloodstream.

Treat Symptoms

While radiation poisoning cannot be reversed with first aid alone, managing symptoms is crucial for stabilizing the person until they can receive professional medical treatment.

Rehydrate: If the person is experiencing vomiting, diarrhea, or dehydration, encourage them to drink clean, uncontaminated water. Oral rehydration solutions (ORS) can help restore electrolytes and prevent dehydration.

Manage pain: Over-the-counter pain relievers like acetaminophen or ibuprofen can help reduce fever, pain, and discomfort. Avoid giving nonsteroidal anti-inflammatory drugs (NSAIDs) if the person has internal bleeding, as these can worsen the condition.

Prevent infections: Since radiation exposure weakens the immune system, it's important to prevent infections. Keep any cuts or burns clean and covered, and monitor for signs of infection such as redness, swelling, or fever.

Limit exposure to contaminants: Encourage the person to rest in a clean, protected environment to minimize further exposure to radioactive particles and infections.

Monitor for Signs of Shock or Infection

Radiation poisoning can lead to serious complications, such as shock or infection, which require immediate attention.

Signs of shock: If the person exhibits symptoms such as pale, cool, clammy skin; rapid breathing; confusion; or a weak pulse, they may be going into shock. Lay them flat, elevate their legs, and keep them warm while seeking emergency medical help.

Signs of infection: Watch for signs of infection, such as high fever, chills, and swelling. If possible, administer antibiotics to prevent or treat infections in individuals with weakened immune systems.

Seek Medical Attention

Radiation poisoning requires professional medical treatment, particularly for individuals exposed to high doses. Treatments such as blood transfusions, antibiotics, and medications to stimulate bone marrow recovery (such as filgrastim) can help manage the effects of radiation poisoning. In severe cases, a stem cell transplant may be needed to restore bone marrow function.

Medications and Treatments for Radiation Poisoning

While first aid is essential for stabilizing a person after radiation exposure, certain medications and treatments can help reduce the severity of radiation poisoning and improve the chances of survival. These include:

Potassium iodide (KI): Potassium iodide can protect the thyroid gland from absorbing radioactive iodine-131, reducing the risk of thyroid cancer. This medication is most effective when taken shortly after exposure to radioactive iodine.

Prussian blue: This medication binds to radioactive cesium and thallium in the intestines, preventing them from being absorbed into the body and allowing them to be excreted.

Filgrastim (Neupogen): This drug stimulates the production of white blood cells and can help restore the immune system after radiation damage to the bone marrow.

Removing the person from the radiation source, decontaminating them, treating symptoms, and preventing complications such as infections or shock are critical steps in managing radiation poisoning. While first aid can provide temporary relief, seeking professional medical treatment is essential for long-term recovery.

The Fallout Shelter: How to Build on a Budget

Building a fallout shelter on a budget is not only a practical step toward ensuring safety in the event of a nuclear disaster, but it can also be accomplished without an enormous financial investment. A well-designed fallout shelter provides critical protection from radiation, blast effects, and potential fallout contamination. Whether you're a survivalist or simply looking to prepare for the worst, creating a shelter that offers adequate protection doesn't have to break the bank. In this chapter, we will explore how to construct a fallout shelter on a budget, focusing on affordable materials, cost-effective design strategies, and essential features to maximize safety while minimizing expenses.

Understanding Fallout Shelter Basics

Before diving into the construction process, it's important to understand what a fallout shelter is and what it needs to achieve. A fallout shelter is designed to protect its occupants from radioactive fallout and, in some cases, the initial blast wave of a nuclear explosion. The key protective factors in a fallout shelter are shielding, distance, and time:

Shielding: The thicker and denser the material between you and the fallout, the better it will block radiation. Materials like concrete, earth, and lead provide excellent shielding.

Distance: The farther you are from the source of radiation, the less exposure you will receive. A fallout shelter should ideally be underground or heavily shielded from the outside environment.

Time: Radioactive fallout loses much of its danger after the first few days or weeks. A fallout shelter needs to provide protection for at least two weeks, as radiation levels drop significantly during this period.

Design Considerations for a Budget-Friendly Shelter

Building a fallout shelter on a budget requires a strategic approach, prioritizing protection while keeping costs low. Here are key design considerations to ensure both safety and affordability.

Location: Utilize Existing Structures

One of the most cost-effective ways to build a fallout shelter is to use or modify an existing structure rather than constructing an entirely new one. Basements, garages, and even crawl spaces can be converted into fallout shelters with relatively minimal expense.

Basements: Basements are naturally suited for fallout shelters because they are already underground, offering some inherent protection from radiation. By reinforcing the walls and ceiling with additional shielding materials (such as concrete, brick, or earth), you can significantly increase the shelter's effectiveness.

Garages or sheds: If you don't have a basement, a detached garage or sturdy shed can be reinforced to serve as a shelter. These structures can be buried with earth or sandbags to provide extra shielding, and the interior can be insulated with protective materials.

Underground crawl spaces: If you have an accessible crawl space beneath your house, you can transform it into a basic fallout shelter by reinforcing the floor above and lining the walls with dense materials like cinder blocks or earth bags.

Excavation: Digging Your Own Shelter

If your property doesn't have an existing structure suitable for conversion, digging an underground shelter is a viable option. An underground fallout shelter provides excellent protection from radiation and blast effects, as the earth itself serves as a natural shield. While professional excavation can be expensive, digging your own shelter is possible with the right tools and planning.

Shovel or rented equipment: If you're working on a budget, manual labor is your most affordable option. Digging a simple trench or small underground room can be accomplished with basic tools like a shovel or, for faster progress, you can rent small excavation equipment.

Earth-covered shelters: Once the excavation is complete, the shelter can be reinforced with affordable materials like wood, corrugated metal, or cinder blocks. Covering the shelter with at least 3 feet of earth will provide excellent radiation shielding and additional protection from the elements.

Materials: Affordable Options for Shielding

The materials you choose for your fallout shelter will have a significant impact on both its protective ability and cost. While high-end materials like steel and lead offer excellent protection, they are often expensive. However, there are plenty of cost-effective alternatives that can still provide adequate shielding from radiation.

Concrete: Concrete is one of the most effective and affordable materials for building a fallout shelter. You can either use precast concrete blocks or pour your own concrete using cement, sand, and gravel. A 6-inch-thick concrete wall provides substantial radiation shielding.

Cinder blocks: Cinder blocks are an inexpensive building material that can be used to create walls and barriers. By stacking cinder blocks and filling them with concrete or sand, you can build strong, protective walls for your shelter.

Sandbags: Sandbags are a highly affordable way to add radiation shielding to your shelter. You can stack sandbags around the walls and roof of an existing structure or use them to line an underground shelter. Sandbags filled with soil or sand provide excellent protection from fallout.

Earth: The earth itself is one of the most effective and budget-friendly shielding materials. If you have access to a large amount of soil or clay, you can bury your shelter in earth to increase radiation protection. A layer of earth at least 3 feet thick is ideal for blocking radiation.

Ventilation: Ensuring Airflow on a Budget

Proper ventilation is essential for any fallout shelter, as occupants will need a continuous supply of fresh air to survive, especially during extended stays. While high-end ventilation systems can be costly, there are affordable ways to ensure adequate airflow.

PVC pipes and vents: Basic PVC pipes or steel ducts can be used to create a ventilation system. Run pipes from the interior of your shelter to the outside, ensuring that the intake and exhaust pipes are positioned far enough apart to prevent contamination of incoming air.

Manual air pumps: For budget shelters, a manual air pump (such as a bellows or hand-cranked fan) can be used to circulate fresh air. Ensure that the air intake is filtered to prevent fallout particles from entering the shelter.

Air filtration: While advanced air filtration systems can be expensive, you can create a basic filtration system using materials like HEPA filters, activated carbon, or even damp cloths to trap fallout particles before they enter the shelter.

Water and Food Storage

A fallout shelter needs to be equipped with enough food and water to last for at least two weeks, as this is the minimum time required for radiation levels to drop to safer levels. Storing enough supplies for your family doesn't have to be costly if you plan carefully.

Water barrels or jugs: Purchase large water barrels or jugs that can store clean drinking water. You can often find affordable options online or at local stores. A general rule is to store 1 gallon of water per person per day, meaning a family of four will need at least 56 gallons of water for two weeks.

Canned and dried food: Stock up on inexpensive, long-lasting food items like canned goods, dried beans, rice, and pasta. These foods are affordable and can be stored for extended periods without refrigeration.

Water purification: In case your stored water supply runs out, have a water purification method ready, such as a water filter or purification tablets. This ensures you can access clean water from outside sources if necessary.

Lighting and Power

While a fallout shelter can function without electricity, having a basic lighting and power system can significantly improve comfort and functionality. For budget shelters, you can rely on low-cost power solutions that don't require a full electrical grid connection.

Battery-powered lights: LED lanterns and flashlights that run on rechargeable batteries or solar power are affordable and provide enough light to navigate the shelter during power outages.

Solar chargers: Small solar-powered chargers can be used to recharge batteries and power small devices like radios and lights. This allows you to maintain communication and basic electronics without relying on external power sources.

Candles or oil lamps: As a low-tech alternative, stockpile candles or oil lamps as a backup light source. However, be cautious with open flames in enclosed spaces and ensure proper ventilation to prevent carbon monoxide buildup.

DIY Fallout Shelter Plans

If you're handy with tools and looking for a step-by-step plan to build your own budget-friendly fallout shelter, here's a basic DIY approach that minimizes costs while maximizing safety.

Step 1: Choose Your Location

Decide whether you'll convert an existing structure (like a basement or shed) or dig an underground shelter. Consider factors like accessibility, proximity to your home, and soil stability.

Step 2: Build the Structure

For underground shelters, excavate a space large enough to accommodate your family and supplies. Reinforce the walls with cinder blocks or wood planks, and cover the shelter with at least 3 feet of earth for radiation protection.

If using an above-ground structure, reinforce the walls with concrete blocks, sandbags, or other dense materials. Focus on shielding the ceiling, as this will be the most exposed area during fallout.

Step 3: Add Ventilation

Install simple PVC pipe vents with filters to provide fresh air. Place one pipe for air intake and one for exhaust, ensuring they are positioned to avoid cross-contamination.

For manual airflow, install a hand-cranked fan or bellows system to circulate air through the shelter.

Step 4: Stock Supplies

Store enough water and non-perishable food to last at least two weeks. Use sturdy, sealed containers to protect food and water from contamination.

Include a basic first aid kit, extra clothing, and sanitation supplies like a portable toilet or waste disposal system.

Step 5: Secure the Entrance

Reinforce the entrance door with thick materials like wood or metal to prevent fallout from entering. Install a second inner door or curtain for added protection against radiation.

If possible, create an "airlock" area just inside the entrance where you can remove contaminated clothing and decontaminate before entering the main shelter.

Cost-Saving Tips for Building a Fallout Shelter

Salvage materials: Reuse materials like wood, bricks, or cinder blocks from other construction projects to reduce costs.

Buy in bulk: Purchase sandbags, concrete, and other bulk materials at wholesale prices to save money.

DIY labor: Doing the excavation and construction work yourself or with family members can dramatically lower the cost of building a fallout shelter.

Local resources: Check for affordable or free resources in your community, such as excess building materials, used furniture, or second-hand supplies.

Building a fallout shelter on a budget is achievable with the right planning, materials, and construction techniques. Whether you're digging an underground shelter or reinforcing a basement, the key to success lies in maximizing your resources and making smart, cost-effective choices.

Choosing the Right Location for Your Shelter

Choosing the right location for your fallout shelter is one of the most important decisions you'll make in your preparation for nuclear disaster survival. The location determines not only how well the shelter will protect you from radiation and fallout but also how accessible and sustainable it will be in the long term. The right site must balance practicality, safety, and affordability. In this chapter, we'll explore the critical factors to consider when choosing a shelter location, including environmental considerations, proximity to your home, geographical features, and accessibility.

Key Factors in Choosing a Shelter Location

When selecting a location for your fallout shelter, it's essential to evaluate how well it meets several critical survival criteria. The location must offer protection from radioactive fallout, be structurally sound, and provide access to essentials like air, water, and food.

Proximity to Your Home

One of the most crucial factors in choosing a shelter location is its proximity to your home. In the event of a nuclear explosion or fallout, time is of the essence, and you'll need to reach your shelter quickly to avoid exposure to dangerous radiation levels. The closer your shelter is to where you live, the better.

Home basements: A basement is a prime location for a fallout shelter because it's already part of your home, offers some natural radiation shielding, and allows you to access it quickly. You can reinforce and adapt the space for additional protection.

Underground structures near your home: If your property includes a garage, cellar, or other structure, these can be reinforced to function as a shelter, provided they are easy to access in an emergency.

Remote shelters: While some people opt for remote shelters (in rural or isolated areas), keep in mind that getting to these locations may not be possible during an emergency, especially if roads are blocked or transportation is compromised.

Underground vs. Above-Ground Shelters

When deciding on the shelter location, consider whether you'll build an underground or above-ground shelter. Each has its advantages, and the best option depends on your environment and available resources.

Underground shelters: These are the most effective in protecting against radiation, as the earth provides natural shielding from fallout. Underground shelters also offer protection from blast waves and extreme temperatures. However, they require more effort to build, including excavation, reinforcement, and waterproofing.

Above-ground shelters: While not as naturally protected as underground shelters, above-ground structures can still be effective if properly reinforced. You can build an above-ground shelter using materials like concrete or cinder blocks and surround it with sandbags or earth for additional radiation shielding. Above-ground shelters are easier to construct, but they may be more vulnerable to blast effects.

Geological Stability

The geological features of the land play an important role in the safety and sustainability of your fallout shelter. If you're building an underground shelter, it's essential to ensure the ground is stable enough to support the structure without risking cave-ins, flooding, or erosion.

Stable soil: Choose a location with firm, stable soil, such as clay or loam. These soil types are less likely to erode or shift, providing a solid foundation for your shelter. Avoid loose or sandy soils, which may be prone to collapsing or shifting.

Water table: Be aware of the water table in your area. If the water table is high, digging an underground shelter may lead to flooding, especially during rainy seasons. Areas with low water tables are ideal for underground shelters as they are less likely to flood.

Slope and drainage: Consider the natural slope of the land. Building on high ground ensures better drainage and reduces the risk of water pooling around your shelter, which could lead to leaks or structural instability. Avoid low-lying areas that may collect water during storms or floods.

Protection from Radiation and Blast Effects

The primary function of a fallout shelter is to protect you from radiation and the harmful effects of a nuclear explosion. When choosing a location, focus on maximizing your protection from both fallout and potential blast waves.

Distance from high-risk areas: Ideally, your shelter should be located far away from potential nuclear targets, such as military bases, power plants, or major cities. The farther you are from these areas, the lower the risk of exposure to dangerous levels of radiation and blast waves.

Natural barriers: If possible, select a location that benefits from natural shielding, such as hills, forests, or rock formations. These features can provide additional protection from radiation and blast waves by absorbing and deflecting energy.

Buried structures: Underground shelters are naturally protected from blast effects, as the earth absorbs much of the shock wave from a nearby explosion. For above-ground shelters, constructing the shelter against a natural barrier (such as a hillside) or using dense materials like concrete and earth can offer additional protection.

Accessibility and Escape Routes

While you want your shelter to be in a protected, secure location, it's also important to consider how accessible it will be during an emergency. In a fallout situation, roads and infrastructure may be damaged or blocked, making it difficult to reach remote or hard-to-access shelters.

Quick access: The shelter should be easy to reach from your home or workplace. Ideally, you should be able to get to the shelter within minutes, especially if radiation exposure is imminent.

Escape routes: In addition to accessibility, consider how you will escape the shelter in case of an emergency, such as structural damage or flooding. Your shelter should have multiple exits or an emergency hatch to allow you to evacuate safely if needed.

Remote locations: If your shelter is located far from home, plan alternative routes and transportation options to ensure you can still reach it if roads are blocked or transportation is disrupted.

Ventilation and Air Supply

Proper ventilation is essential for a fallout shelter, especially in an underground or enclosed space where fresh air may be limited. The shelter must be able to provide a continuous supply of clean air, free from fallout particles, to ensure the health and safety of its occupants.

Air intake and exhaust: Install air intake and exhaust vents that allow fresh air to circulate through the shelter. These should be placed at a safe distance from one another to prevent contaminated air from entering the shelter.

Filtered air: Ensure that the air intake is equipped with filters, such as HEPA filters or activated carbon, to remove fallout particles and other contaminants from the air before it enters the shelter.

Manual ventilation systems: In a budget shelter, manual air pumps or hand-cranked ventilation systems can provide air circulation if power is unavailable.

Water and Food Sources

A long-term fallout shelter requires access to clean water and food supplies to sustain you for an extended period. When choosing a location, consider how you will store and access these essential resources.

Water access: Ideally, your shelter should be located near a clean water source, such as a well, stream, or underground aquifer. However, in a fallout situation, it's essential to store large quantities of water inside the shelter, as external water sources may be contaminated by radioactive particles.

Food storage: Your shelter should have enough space to store at least two weeks' worth of food for each occupant. Choose a location with enough room for airtight containers or shelves to store canned and dried foods, as well as supplies like vitamins and first aid items.

Secondary sources: Consider how you will replenish your food and water supplies if you need to stay in the shelter for an extended period. If possible, plan for alternative water sources (such as rainwater collection) or the ability to grow food in a controlled, indoor environment.

Ideal Locations for a Fallout Shelter

With all the factors discussed, certain locations make better fallout shelters than others. Here are a few examples of ideal locations for constructing or converting a fallout shelter on a budget:

Basement Shelters

Your home's basement is one of the most convenient and affordable locations for a fallout shelter. It's already partially underground, providing natural protection from radiation, and it allows you to access the shelter quickly in an emergency. You can enhance its protective ability by reinforcing the walls with concrete or sandbags and installing a ventilation system to ensure fresh air.

Backyard Underground Shelters

If you have the space, building an underground shelter in your backyard is an excellent option. By digging into the earth and reinforcing the structure with concrete, cinder blocks, or earthbags, you can create a highly effective fallout shelter. Underground shelters provide the best protection from both radiation and blast effects. Consider covering the shelter with at least 3 feet of earth for optimal shielding.

Converted Garages or Sheds

If you don't have a basement or the ability to dig an underground shelter, consider converting a garage or shed into a fallout shelter. Reinforce the structure with additional layers of concrete, brick, or sandbags, and create an inner shelter room with ventilation, air filtration, and a separate entrance. This is a cost-effective solution that provides reasonable protection while utilizing an existing structure.

Natural Caves or Rock Shelters

Natural caves and rock shelters can offer excellent protection from radiation and blast waves due to the density and thickness of the surrounding rock. If you live in an area with natural cave systems, you may be able to adapt part of the cave into a fallout shelter by installing ventilation, lighting, and supplies. Keep in mind that natural shelters may require additional reinforcement to ensure structural stability.

Remote Underground Bunkers

For those with the resources and land to do so, building a dedicated underground bunker in a remote area is an ideal solution. While this can be more expensive than converting an existing structure, it offers the highest level of protection. A remote underground bunker can be designed with long-term survival in mind, including ample space for food, water, ventilation systems, and even renewable power sources like solar panels.

Choosing the right location for your fallout shelter is crucial for maximizing protection and ensuring your long-term survival. Whether you're converting an existing basement, digging an underground shelter, or constructing a reinforced above-ground structure, the location should offer easy access, natural shielding, and protection from radiation and blast effects. Consider factors such as proximity to your home, geological stability, ventilation, and access to essential resources like food and water. With careful planning, you can build a safe, effective fallout shelter that provides you and your family with the protection you need in a post-nuclear environment.

Materials for Shielding: What Works and Why

Building a fallout shelter that effectively shields against radiation requires using materials that block or reduce the amount of ionizing radiation reaching the occupants. The materials you choose for shielding are critical to the shelter's protective capability. Different materials offer varying levels of protection based on their density and thickness, which affects how well they absorb or deflect radiation. In this chapter, we will explore which materials work best for shielding, why they are effective, and how to use them cost-effectively in building a fallout shelter.

Understanding Radiation and Shielding

To understand why certain materials work better for shielding, it's important to first understand the types of radiation emitted during a nuclear explosion or from fallout. There are three main types of ionizing radiation that fallout shelters must protect against:

Alpha radiation: These are heavy, charged particles that can be stopped by thin materials like paper or even skin. While alpha particles are not a threat externally, they become dangerous if inhaled or ingested.

Beta radiation: These are fast-moving, smaller particles that can penetrate the outer layer of skin but are easily blocked by materials like plastic, glass, or a few millimeters of aluminum.

Gamma radiation: The most dangerous type of radiation in terms of penetration, gamma rays are highly energetic and can pass through most materials. Shielding against gamma radiation requires dense, thick materials like lead, concrete, or earth.

The goal of effective radiation shielding is to absorb or deflect as much radiation as possible before it reaches the shelter's occupants. The thicker and denser the material, the better it will protect against gamma radiation, the primary threat during fallout.

Key Materials for Shielding and Why They Work

Various materials are commonly used for fallout shelter construction, each with unique properties that make them effective for blocking radiation. The effectiveness of a material is determined by its density (how much mass it has in a given volume) and thickness (how much material is present to absorb radiation).

Concrete

Why it works: Concrete is one of the most widely used materials for fallout shelter construction because of its density, affordability, and availability. Concrete is highly effective at absorbing gamma radiation due to its dense structure. A typical 12-inch thick concrete wall can reduce radiation exposure to about 10% of the outside level.

How to use it: Concrete can be poured into forms to create walls, floors, and ceilings, or precast concrete blocks can be used to build the shelter. Reinforcing concrete with steel bars (rebar) adds strength, making it ideal for underground or blast-resistant shelters.

Recommended thickness: A 3-foot (1 meter) thick concrete wall provides substantial protection from gamma radiation, but even thinner walls (12-24 inches) can still offer significant shielding.

Lead

Why it works: Lead is one of the most effective materials for blocking radiation due to its high density. Just a few millimeters of lead can block a significant amount of gamma radiation, making it a popular choice for radiation shielding in medical and industrial applications.

How to use it: While lead is highly effective, it is also expensive and heavy. It's best used in situations where space is limited, and you need maximum shielding in a small area. Lead sheets or plates can be installed within walls or around doors and windows to enhance radiation protection in critical areas.

Recommended thickness: A lead shield of 1 to 2 inches (2.5 to 5 cm) is typically sufficient to block most gamma radiation, though even thinner layers offer considerable protection.

Earth

Why it works: Earth (soil or dirt) is a naturally abundant and cost-effective material that provides excellent protection against radiation. The density of earth varies depending on its composition (clay, sand, loam), but it offers significant shielding, especially when used in thick layers. Earth absorbs and blocks gamma radiation, making it ideal for covering underground or semi-buried shelters.

How to use it: Earth can be used to cover the roof and sides of an underground or partially underground shelter. It can also be piled around above-ground structures to create earthen walls. The key to using earth effectively is ensuring that it is compacted and that the shelter structure is sturdy enough to support the weight.

Recommended thickness: A layer of 3 feet (1 meter) of compacted earth provides good radiation protection. For increased safety, using 6 feet (2 meters) of earth offers superior shielding.

Sandbags

Why it works: Sandbags filled with sand or soil are an easy and affordable way to add shielding to a shelter. The individual sand particles absorb and scatter radiation, reducing the overall exposure to gamma rays. Sandbags can be stacked to create thick, dense walls that provide substantial protection.

How to use it: Sandbags are especially useful for reinforcing existing structures or as a temporary shielding solution. They can be stacked on the sides and roof of a shelter, or even used to fill gaps in walls. Sandbags are also versatile for creating protective barriers in an emergency.

Recommended thickness: A sandbag wall of 2 to 3 feet (60-90 cm) thick provides significant protection from gamma radiation. For better protection, stacking the bags to a thickness of 4 feet (120 cm) or more is recommended.

Water

Why it works: Water is surprisingly effective at blocking radiation, particularly gamma rays, due to its density and hydrogen content. Water absorbs radiation and disperses its energy, making it a useful material for both permanent and temporary shelters.

How to use it: Water can be stored in large containers, such as drums, jugs, or barrels, and placed along the walls or ceiling of a shelter to increase shielding. In an emergency, a large amount of water can be used to create makeshift barriers around a shelter.

Recommended thickness: A layer of 3 feet (1 meter) of water can block about 50% of gamma radiation. If space permits, thicker layers offer better protection.

Cinder Blocks

Why it works: Cinder blocks, also known as concrete masonry units, are a lightweight alternative to solid concrete. They are commonly used in construction and provide decent radiation shielding when filled with concrete or sand. Though not as dense as solid concrete, cinder blocks are an affordable option for building radiation-resistant structures.

How to use it: Cinder blocks can be stacked to create walls, with the hollow spaces filled with sand, gravel, or concrete to increase their density. They are ideal for constructing above-ground shelters or reinforcing existing buildings.

Recommended thickness: A cinder block wall filled with concrete or sand, and at least 12 inches (30 cm) thick, can offer moderate radiation protection. For better shielding, aim for a wall thickness of 24 to 36 inches (60 to 90 cm).

Brick

Why it works: Brick is a relatively dense building material that provides good protection against radiation when used in thick walls. While not as dense as concrete, brick offers a cost-effective option for building or reinforcing shelters.

How to use it: Brick walls can be used to construct above-ground shelters or to reinforce basement or garage walls. Adding a second layer of bricks or combining bricks with other materials like concrete or sandbags enhances the overall shielding.

Recommended thickness: A double-layered brick wall (24 inches or 60 cm thick) can provide substantial radiation protection. For more effective shielding, adding a second layer of bricks or reinforcing with concrete is recommended.

Steel

Why it works: Steel is a strong, dense material that provides good protection against radiation when used in thick sheets or panels. It is also resistant to blast effects, making it useful for shelters that may be exposed to explosions or impacts.

How to use it: Steel plates or sheets can be incorporated into the walls, ceiling, or doors of a shelter. Steel is often used in conjunction with other materials, such as concrete or earth, to provide a combination of strength and radiation protection.

Recommended thickness: A steel plate of 0.5 to 1 inch (1.3 to 2.5 cm) thick provides good gamma radiation shielding. Thicker plates can be used for areas that require additional protection.

Building an Effective Shield with Combined Materials

Using a combination of materials is often the most practical and effective way to create a fallout shelter that offers strong protection without excessive cost. For example:

Concrete and earth: Build the core structure of the shelter with concrete or cinder blocks and cover it with a thick layer of earth for added shielding.

Sandbags and water: Stack sandbags around the shelter walls and roof and place barrels of water inside for dual protection.

Lead and concrete: Use lead sheeting in high-exposure areas like doors or windows, while constructing the main walls from concrete for overall shielding.

Thickness and Density: The Key to Protection

When it comes to radiation shielding, both the density and the thickness of the material are critical. For example, a thinner layer of lead may provide better protection than a thicker layer of less dense material like wood or plastic. However, if dense materials like lead or steel are too expensive or impractical, layering less dense materials like earth, sandbags, or concrete can still provide excellent protection when used in sufficient thickness.

In general:

Doubling the thickness of a shielding material reduces radiation exposure by about half.

Thicker materials like earth and concrete should be used for the shelter's walls and roof, while denser materials like lead or steel can be used in high-exposure areas for enhanced protection.

Cost-Effective Shielding Solutions

While lead and thick concrete provide the best protection, they can be expensive and heavy to work with. For budget-conscious shelter builders, here are some cost-effective alternatives:

Use local earth: Instead of purchasing expensive materials, use the earth from your property to cover your shelter. Compacting the soil around an underground or partially buried shelter is a highly effective way to create inexpensive shielding.

Salvage materials: Reuse bricks, cinder blocks, or concrete from demolition sites or construction projects. These materials are often available at low cost or even free and can be used to build sturdy shelter walls.

DIY sandbags: Instead of purchasing pre-filled sandbags, buy inexpensive burlap or plastic bags and fill them with sand or soil from your property. Sandbags are versatile and can be stacked to provide both temporary and long-term protection.

Choosing the right materials for shielding your fallout shelter is essential for ensuring the safety and protection of its occupants. Dense, thick materials like concrete, lead, and earth offer the best protection against gamma radiation, but combining different materials can create a cost-effective and highly protective shelter. By carefully considering the materials you have access to and using affordable options like sandbags, cinder blocks, and water, you can build a robust fallout shelter that will keep you safe in a nuclear disaster.

Ventilation Systems: Keeping Fallout Out and Air In

Ventilation is one of the most critical aspects of designing a fallout shelter, as it ensures that fresh air flows into the shelter while keeping radioactive particles and harmful contaminants out. Proper ventilation allows you and your family to breathe safely inside the shelter, even during extended stays, while preventing the buildup of carbon dioxide and other hazardous gases. In this chapter, we'll cover how to design an effective ventilation system on a budget, the types of air filtration needed to block radioactive fallout, and how to maintain proper airflow in a sealed environment.

Why Ventilation is Essential in a Fallout Shelter

In a fallout shelter, maintaining a fresh supply of clean air is just as important as shielding from radiation. Without proper ventilation, the air inside the shelter can quickly become stale, making it difficult to breathe and leading to dangerous carbon dioxide buildup. Additionally, radioactive fallout particles from the outside environment can enter the shelter through the air intake, posing a serious health risk if not properly filtered.

To ensure your shelter provides safe, breathable air, a ventilation system must meet three key requirements:

Fresh air intake: The shelter must have a consistent supply of fresh air from the outside to replace stale air.

Air filtration: The air entering the shelter must be filtered to remove fallout particles, dust, and other contaminants.

Air circulation: Inside the shelter, the air needs to be circulated to prevent pockets of carbon dioxide from building up and to keep the air fresh and breathable.

Types of Ventilation Systems

Ventilation systems in fallout shelters can range from simple, low-tech designs to advanced air filtration systems. The best option for your shelter depends on the level of protection you need, your budget, and the materials available. Here are the main types of ventilation systems to consider:

Passive Ventilation (Natural Airflow)

What it is: Passive ventilation systems rely on natural airflow to bring fresh air into the shelter and exhaust stale air. This can be achieved by placing intake and exhaust vents in strategic locations to allow air to flow naturally through the shelter. Passive systems are simple and inexpensive but may not be as effective in controlling airflow or filtering contaminants.

How it works: Passive ventilation systems use the difference in air pressure and temperature between the inside and outside of the shelter to create airflow. For example, by positioning the intake vent low to the ground and the exhaust vent higher up, you can take advantage of the natural flow of air as cooler air enters from below and warmer air rises and exits.

Pros: Passive ventilation is inexpensive, easy to set up, and requires no electricity. It works well in mild climates and with low-tech shelters.

Cons: Passive systems don't offer much control over airflow and may not provide sufficient filtration. They are also less effective in extreme conditions or sealed environments.

Manual Ventilation (Hand-Cranked Systems)

What it is: Manual ventilation systems use hand-cranked devices to move air through the shelter. These systems are an affordable option for shelters without access to electricity and are ideal for situations where you need to ensure a steady flow of air into and out of the shelter.

How it works: A hand-cranked fan or bellow is connected to an air intake or exhaust vent. When the crank is turned, fresh air is pulled into the shelter through the intake vent, and stale air is expelled through the exhaust vent. Manual ventilation systems can also include basic filters to prevent fallout particles from entering the shelter.

Pros: Manual systems are inexpensive, require no electricity, and provide reliable airflow. They are also easy to install and operate.

Cons: These systems require human effort to operate and may not provide continuous airflow unless someone is cranking the device regularly. They also offer limited air filtration.

Powered Ventilation (Electric Fans and Blowers)

What it is: Powered ventilation systems use electric fans or blowers to circulate air through the shelter. These systems provide more control over airflow and can be equipped with advanced filters to block fallout particles. Powered ventilation is ideal for more permanent shelters or situations where consistent airflow is essential.

How it works: Electric fans are installed in both the air intake and exhaust vents. The intake fan pulls fresh air into the shelter, while the exhaust fan expels stale air. Powered ventilation systems often include HEPA filters, carbon filters, or other types of filtration to ensure the air entering the shelter is free from radioactive particles and other contaminants.

Pros: Powered systems offer more control over airflow, better filtration, and consistent air circulation. They are effective for both long-term shelters and short-term emergency use.

Cons: These systems require a power source, such as a generator, battery, or solar panels. Without a reliable source of electricity, powered ventilation may not be sustainable during extended shelter stays.

Air Filtration Systems: Keeping Fallout Out

In addition to ensuring proper airflow, a ventilation system must include air filtration to block radioactive fallout particles from entering the shelter. Fallout particles are extremely dangerous when inhaled, as they can lodge in the lungs and expose the body to high levels of radiation. Therefore, your ventilation system needs to have an air filtration component to keep the air inside the shelter safe to breathe.

HEPA Filters

What it is: High-efficiency particulate air (HEPA) filters are designed to trap fine particles, including radioactive dust and fallout. These filters are highly effective at removing particles as small as 0.3 microns, making them an excellent choice for fallout protection.

How it works: HEPA filters are placed in the air intake of the ventilation system. As air passes through the filter, fallout particles are trapped in the dense mesh of fibers, allowing only clean air to enter the shelter. HEPA filters are often combined with other filtration methods, such as activated carbon filters, to remove both particles and gases.

Pros: HEPA filters are highly effective at blocking fallout particles and other contaminants. They are readily available and can be used in both manual and powered ventilation systems.

Cons: HEPA filters need to be replaced periodically, as they become clogged with particles. They can also reduce airflow if not properly maintained.

Activated Carbon Filters

What it is: Activated carbon filters are designed to remove gases, odors, and chemicals from the air. While they are not as effective at blocking particles as HEPA filters, activated carbon filters are excellent for removing airborne toxins and improving air quality in a sealed environment.

How it works: Activated carbon filters use porous carbon to absorb gases and chemicals from the air. When air passes through the filter, harmful substances are trapped in the carbon's pores, leaving clean, odor-free air. These filters are often used in combination with particle filters to provide comprehensive air filtration.

Pros: Activated carbon filters are effective at removing gases and chemicals that may enter the shelter along with fallout particles. They help improve overall air quality and reduce odors.

Cons: Carbon filters need to be replaced periodically, as they lose effectiveness over time. They also do not filter out particles like fallout dust, so they must be used alongside HEPA or other particle filters.

Pre-Filters

What it is: Pre-filters are coarse filters designed to capture large particles, such as dust, debris, and larger fallout particles, before they reach the primary filter. These filters help extend the life of more expensive HEPA or activated carbon filters by reducing the amount of contamination they need to handle.

How it works: Pre-filters are placed in front of the main filter in the air intake. As air flows through the system, larger particles are trapped by the pre-filter, allowing cleaner air to pass through to the primary filter. Pre-filters are typically made of materials like foam, mesh, or fabric.

Pros: Pre-filters are inexpensive and help protect the main filters from clogging. They are easy to replace and reduce maintenance costs.

Cons: Pre-filters are not effective at blocking fine particles, so they should always be used in conjunction with a HEPA filter or other fine particle filters.

DIY Air Filtration Solutions

If you're building a shelter on a budget, you can create simple DIY air filters using readily available materials. While not as effective as commercial filters, these DIY solutions can still help reduce fallout particles and improve air quality in an emergency.

Wet cloth or sponge filters: A basic method for filtering fallout particles involves placing a wet cloth or sponge over the air intake. The water traps radioactive dust and particles, preventing them from entering the shelter. This method provides temporary filtration and should be used with other ventilation systems for more effective results.

Homemade carbon filters: You can create a basic carbon filter using activated charcoal (available at pet stores or online) placed inside a filter box or container. Air drawn through the charcoal will be stripped of gases and some particles, improving air quality in the shelter.

Ensuring Proper Air Circulation Inside the Shelter

Once fresh, filtered air enters the shelter, it's essential to maintain proper air circulation to prevent carbon dioxide buildup and ensure that stale air is continuously replaced with fresh air. Here's how to achieve good air circulation inside the shelter:

Use Multiple Air Vents

Your shelter should have at least two vents: one for air intake and one for air exhaust. Position the air intake low on one wall and the exhaust vent high on the opposite wall to create natural airflow through the shelter. This ensures that fresh air enters at ground level while stale air exits near the ceiling, where warm air and carbon dioxide accumulate.

Manual Air Circulation Devices

In a budget shelter, simple devices like hand-cranked fans or bellows can help circulate air inside the shelter. These manual devices move air through the vents and prevent the air from becoming stagnant. Regularly using these devices ensures a consistent exchange of fresh air and minimizes the risk of carbon dioxide buildup.

Powered Air Circulation

If your shelter has access to a power source, you can use small electric fans to maintain air circulation. These fans can be placed strategically to move air from the intake vent across the shelter and toward the exhaust vent. Powered air circulation is especially useful in larger shelters where airflow may be more difficult to control.

Managing Carbon Dioxide Buildup

Even with proper ventilation, carbon dioxide (CO_2) can accumulate in a sealed shelter, particularly during long stays. High levels of CO_2 can cause dizziness, headaches, and even suffocation if not managed properly. Here are some ways to prevent CO_2 buildup:

Air circulation: Ensure that fresh air is continuously flowing through the shelter by using manual or powered ventilation systems.

CO_2 absorbers: In situations where ventilation is limited, CO_2 absorbers, such as soda lime or calcium hydroxide, can be used to remove excess CO_2 from the air inside the shelter. These materials chemically react with CO_2 to absorb it, helping maintain safe air quality.

Regular ventilation checks: Monitor the shelter's ventilation system to ensure that air is circulating properly and that vents are not clogged with debris or fallout particles. If necessary, adjust the system to increase airflow and improve air exchange.

Ventilation is a vital component of any fallout shelter, ensuring that fresh air enters the shelter while keeping radioactive fallout particles out. Whether you choose a simple passive system, a manual hand-cranked setup, or a powered ventilation system with advanced filtration, it's essential to ensure that your shelter maintains a consistent flow of clean air. By incorporating HEPA filters, activated carbon filters, and pre-filters into your ventilation system, you can effectively block fallout particles and improve air quality inside the shelter. Proper air circulation and CO_2 management will help keep the air safe and breathable, ensuring that your fallout shelter provides a secure environment during a nuclear event.

Creating an Underground Haven: Building Below Ground

Creating an underground fallout shelter is one of the most effective ways to protect yourself and your family from the dangers of nuclear fallout, radiation, and other catastrophic events. The earth itself acts as a natural barrier, providing excellent protection against radiation, extreme temperatures, and blast waves from nuclear explosions. Building a below-ground haven can be a complex and labor-intensive process, but with proper planning and the right approach, you can create a durable and safe underground shelter. In this chapter, we'll cover the key steps to constructing an underground shelter, including site selection, excavation, structural design, and ensuring long-term sustainability.

Why Build Underground?

There are several key reasons why an underground shelter offers the best protection during a nuclear event:

Radiation shielding: The earth provides significant protection against gamma radiation, the most dangerous type of radiation emitted after a nuclear explosion. Even a few feet of soil can block most of the harmful rays, making an underground shelter one of the safest places to be during fallout.

Blast protection: Underground shelters are naturally insulated from the shock waves of nuclear explosions. The surrounding earth absorbs much of the force, reducing the risk of structural damage or injury from debris.

Temperature control: Underground shelters remain relatively cool in the summer and warm in the winter, making them more comfortable and energy-efficient than above-ground structures.

Concealment: An underground shelter is less visible than an above-ground structure, providing an extra layer of security in a post-disaster environment where looting or social unrest may occur.

Key Considerations for Building an Underground Shelter

Before you begin construction, it's important to plan carefully and consider several critical factors that will influence the success of your underground shelter.

Site Selection

Choosing the right location for your underground shelter is crucial. The site should provide natural protection, be geologically stable, and have good drainage to avoid flooding.

High ground: Select a location on slightly elevated terrain to prevent water from pooling or flooding the shelter. Avoid low-lying areas or spots near bodies of water.

Distance from your home: Ideally, your underground shelter should be close enough to your home to allow for quick access in an emergency, but far enough away to minimize the risk of structural damage if your home is affected by a nuclear blast or fire.

Geological stability: The soil type and ground conditions will impact both the ease of excavation and the stability of the shelter. Look for firm, stable soils like clay or loam, which provide good support and minimize the risk of shifting or erosion. Avoid loose or sandy soils, which may be prone to collapse.

Excavation

Excavating the site is one of the most challenging parts of building an underground shelter, but it's a necessary step to ensure you have enough depth to provide adequate protection from radiation and blast effects. Here are the main excavation methods and considerations:

Manual excavation: If you're on a tight budget, you can dig the shelter manually using shovels, picks, and wheelbarrows. While this method is labor-intensive and time-consuming, it can be a viable option for smaller shelters.

Mechanical excavation: For larger projects, renting excavation equipment such as backhoes or excavators can speed up the digging process significantly. While more expensive, mechanical excavation is essential for deeper or larger shelters.

Depth: The depth of your shelter is critical for radiation protection. A minimum depth of 10 to 12 feet below ground level is recommended to provide sufficient shielding from gamma rays. Deeper shelters offer better protection but may require more extensive excavation and reinforcement.

Structural Design and Materials

Once the site is excavated, the next step is designing and building the structure of the shelter. The primary goal is to create a durable, reinforced space that can withstand the pressures of the surrounding earth while providing adequate radiation shielding and safety.

Walls and roofing: The walls and ceiling of the shelter should be built using strong, durable materials like reinforced concrete, cinder blocks, or steel. These materials provide excellent structural support and radiation shielding. Concrete is especially effective because of its density and ability to absorb gamma radiation.

Concrete: A 12-inch-thick concrete wall provides good protection from radiation and supports the weight of the earth above.

Cinder blocks: These can be used for wall construction if filled with concrete or sand for added strength and radiation protection.

Steel beams: Steel beams can be used to reinforce the roof and walls, particularly in areas prone to seismic activity or heavy soil loads.

Waterproofing: Proper waterproofing is essential for preventing leaks and moisture buildup inside the shelter. Use waterproof membranes, coatings, or sealants to protect the walls and roof. Additionally, consider installing a drainage system to direct groundwater away from the shelter.

Entrance and exits: Your shelter should have at least one reinforced entrance, preferably with an airtight, blast-resistant door. Consider creating a secondary escape route, such as a hidden hatch or exit tunnel, in case the main entrance becomes blocked or damaged.

Ventilation and Air Filtration

Ventilation is critical in an underground shelter, as the confined space can quickly become unbreathable without a steady supply of fresh air. Your ventilation system should include both air intake and exhaust vents, along with air filtration to block radioactive fallout particles.

Air intake and exhaust: Position the intake vent at a low point and the exhaust vent at a higher elevation to create natural airflow. Use PVC pipes or steel ducts for ventilation channels, and ensure that vents are shielded from fallout and debris.

Manual or powered ventilation: For small shelters, a hand-cranked ventilation system can circulate air. Larger shelters may benefit from electric fans or blowers connected to a generator or solar power system.

Air filtration: Install HEPA filters or activated carbon filters in the air intake vent to remove fallout particles, dust, and other contaminants from the incoming air.

Water and Drainage Management

Water is a critical factor when building underground. You need to prevent flooding and ensure a reliable supply of drinking water while avoiding contaminated water sources.

Drainage system: Install a drainage system around the shelter's foundation to prevent water from seeping in. Perforated drainage pipes (also known as French drains) can direct groundwater away from the shelter and into a safe drainage area.

Sump pump: A sump pump is a good addition for shelters in areas with high water tables. The pump removes any water that collects at the lowest point of the shelter.

Water storage: Ensure that your shelter has an adequate supply of clean water, stored in large containers or barrels. A general rule is to store 1 gallon of water per person per day for at least two weeks.

Insulation and Temperature Control

Underground shelters tend to have stable temperatures, but you'll still need to ensure the space is insulated to maintain comfort during both hot and cold weather. Insulation can also help regulate humidity levels inside the shelter.

Thermal insulation: Use foam boards, spray foam, or fiberglass insulation to line the interior walls and ceiling. This will help keep the shelter cool in summer and warm in winter.

Ventilation for temperature control: A proper ventilation system not only keeps the air fresh but also helps regulate temperature by circulating warm or cool air.

Power Supply

A reliable power source is essential for lighting, ventilation, and powering essential equipment in your shelter. Depending on your budget and needs, there are several options for powering your underground haven:

Generators: Gasoline, diesel, or propane-powered generators are reliable for short-term use but require a steady fuel supply. Ensure the generator is placed outside the shelter to avoid carbon monoxide buildup.

Solar power: Solar panels can provide a renewable energy source for shelters. Store the power in batteries to provide electricity during the night or on cloudy days.

Battery systems: For temporary or emergency power, consider storing a supply of rechargeable batteries to power lights, fans, and communication devices.

Long-Term Sustainability

To make your underground haven viable for extended stays, you'll need to plan for long-term sustainability. This includes storing enough food and water, managing waste, and ensuring mental and physical well-being during an extended shelter stay.

Food storage: Stockpile non-perishable, shelf-stable foods like canned goods, freeze-dried meals, and bulk grains. Consider building shelving or storage areas within the shelter to keep food organized and protected from pests.

Sanitation and waste management: Install a portable toilet or chemical toilet for waste management, and store enough sanitation supplies (such as toilet paper, disinfectant, and trash bags) to last several weeks.

Entertainment and comfort: To ensure mental well-being, stock the shelter with items for entertainment, such as books, games, or radios. Comfortable bedding, extra clothing, and blankets are also essential for staying warm and comfortable.

Step-by-Step Process for Building an Underground Shelter

Step 1: Choose Your Location

Assess your property and select a location that is geologically stable, well-drained, and accessible.

Step 2: Excavate the Site

Dig the shelter to the desired depth, aiming for at least 10 to 12 feet below ground level.

Step 3: Construct the Shelter Structure

Build the walls, floor, and ceiling using reinforced concrete, cinder blocks, or steel.

Ensure that the structure is watertight and able to withstand the pressure of the surrounding earth.

Step 4: Install Ventilation and Air Filtration

Set up air intake and exhaust vents, along with HEPA or carbon filters to ensure clean air.

Step 5: Waterproof the Shelter

Apply waterproof coatings or membranes to the exterior walls and install drainage systems to keep water out.

Step 6: Set Up Water and Power Systems

Install water storage containers and backup power systems like generators or solar panels.

Step 7: Prepare the Interior

Insulate the shelter to maintain comfortable temperatures, and set up storage areas for food, water, and other supplies.

Step 8: Test and Maintain the Shelter

Test the ventilation, power, and drainage systems regularly to ensure the shelter remains safe and functional.

Building an underground fallout shelter provides unparalleled protection from radiation, nuclear blasts, and extreme weather. While it requires significant planning, excavation, and structural reinforcement, the benefits of safety and security in a post-nuclear environment are well worth the effort. By carefully selecting a site, using durable materials, and ensuring proper ventilation, water management, and sustainability, you can create an underground haven that will keep you and your family safe in the most extreme conditions.

Above-Ground Shelters: Can They Work?

While underground shelters are often considered the gold standard for protection during a nuclear event, building an above-ground fallout shelter can be a viable alternative if an underground option is not feasible. Above-ground shelters can be constructed more quickly, are often less expensive, and provide effective radiation protection if built properly. Although they may not offer the same level of shielding as underground shelters, with the right design and materials, an above-ground fallout shelter can still be a safe, practical solution for many scenarios.

In this chapter, we'll explore whether above-ground shelters can work in a post-nuclear world, the essential design elements needed to ensure safety, and how to build one on a budget. We'll also discuss strategies for maximizing radiation protection and blast resistance for above-ground shelters.

Can Above-Ground Shelters Offer Adequate Protection?

The primary challenge of an above-ground shelter is providing sufficient shielding from radiation and, to a lesser extent, protecting against the shockwaves and debris from a nuclear explosion. However, with the right materials and thickness, an above-ground shelter can effectively reduce radiation exposure to a safe level. The key to success lies in using dense, protective materials and ensuring the structure is reinforced to withstand the environment in a fallout zone.

Key Factors for Building an Effective Above-Ground Shelter

When building an above-ground fallout shelter, several factors must be considered to ensure it provides adequate protection. These include the choice of materials, structural design, radiation shielding, and blast resistance.

Radiation Shielding

Radiation shielding is the most critical aspect of any fallout shelter, especially for above-ground structures. Because above-ground shelters are exposed to more fallout particles, you need to use dense materials to block harmful gamma radiation. Effective shielding depends on both the thickness and density of the materials you use.

Concrete: As with underground shelters, concrete is one of the best materials for blocking radiation. For an above-ground shelter, walls should be at least 12 inches thick to provide sufficient gamma radiation shielding. The roof must also be reinforced with concrete or another dense material to block radiation from above.

Cinder blocks: Cinder blocks are a cost-effective alternative to solid concrete. When filled with sand or concrete, they offer substantial protection. A double layer of cinder blocks, filled with sand, can provide strong radiation shielding for an above-ground shelter.

Sandbags: Sandbags filled with dirt or sand can be stacked to create radiation-resistant walls. Sand is an effective barrier against gamma rays, and stacking sandbags to a thickness of 2-4 feet provides a good level of protection. Sandbags are especially useful for reinforcing existing structures or creating temporary fallout shelters.

Earth berms: Building an earth berm (a mound of soil or dirt) around the exterior of your shelter adds another layer of radiation shielding. By covering the lower half or entire exterior of the structure with compacted earth, you can increase the overall protection. A berm of at least 3 feet in height and thickness can significantly reduce radiation exposure.

Lead: Lead is one of the most effective materials for blocking radiation due to its high density. While lead is expensive, using lead sheeting or panels in critical areas (such as doors, windows, or thinner walls) can provide superior shielding.

Structural Integrity

Above-ground shelters need to be structurally sound to withstand environmental pressures, such as wind, debris, and potentially even the shockwaves from a nuclear explosion. Reinforcing the structure is crucial for ensuring it doesn't collapse or fail in a disaster.

Reinforced walls: To resist the force of a blast, use reinforced concrete or cinder block walls with rebar or steel mesh inside. This reinforcement increases the structural integrity of the shelter, helping it withstand external pressures.

Blast-resistant doors and windows: The doors and windows are the weakest points of any shelter. For an above-ground structure, use steel or lead-lined doors designed to withstand both radiation and the pressure of a blast wave. Windows should be avoided, but if they are necessary, use shatter-resistant glass or seal them with heavy coverings like steel plates or leaded glass.

Roof reinforcement: The roof must be strong enough to handle not only the weight of the shielding materials but also any debris that might fall on it. Use concrete, metal, or sandbags to reinforce the roof. Additionally, adding a slope to the roof can help deflect debris and prevent accumulation.

Ventilation and Air Filtration

Like any fallout shelter, an above-ground shelter requires a ventilation system to bring in fresh air and remove carbon dioxide. However, the system must also filter out fallout particles to prevent radiation from entering the shelter.

Filtered air intake: Install an air intake vent that includes HEPA filters or activated carbon filters to remove radioactive dust and contaminants from the incoming air. The intake should be placed away from the direct path of fallout or shielded by earth or other barriers.

Exhaust vent: A high exhaust vent is necessary to expel stale air and carbon dioxide. Make sure the intake and exhaust vents are positioned on opposite sides of the shelter to ensure a steady airflow.

Manual or powered ventilation: In the absence of power, manual ventilation systems, such as hand-cranked fans or bellows, can keep air circulating. If power is available, electric fans and blowers can provide more consistent ventilation, but ensure there's a backup plan if electricity is lost.

Blast Resistance

While an above-ground shelter may not be as naturally protected from blast waves as an underground structure, certain design features can improve its resistance to shockwaves and falling debris.

Reinforced corners: Reinforce the corners and joints of the shelter to reduce the risk of collapse. Using steel reinforcements or thick concrete can help the structure withstand the force of a nearby blast.

Debris deflection: A sloped roof or rounded design can help deflect debris and shockwaves, reducing the impact on the structure. Avoid flat roofs, as they are more likely to collapse under the weight of debris.

Earth berming: Building up earth around the walls of the shelter not only provides radiation shielding but also helps absorb some of the energy from a shockwave. The earth acts as a cushion, protecting the walls from the full force of a nearby explosion.

Building an Above-Ground Shelter: Step-by-Step Guide

Step 1: Choose a Location

Select a site that is easily accessible but away from potential hazards such as power plants, major cities, or military targets.

Ideally, the site should have some natural protection, such as hills or trees, that can help shield the shelter from radiation and blast effects.

Step 2: Excavate and Build the Foundation

Dig a shallow foundation (at least 2 feet deep) for the shelter. This adds stability and allows you to build thicker, reinforced walls.

Lay a reinforced concrete slab as the foundation to support the weight of the shelter and provide a stable base.

Step 3: Construct the Walls

Use dense materials like concrete, cinder blocks filled with sand, or sandbags for the walls. Ensure the walls are at least 12 inches thick for concrete or 2-4 feet thick for sandbags.

Reinforce the walls with steel bars or rebar to increase structural integrity and improve blast resistance.

Step 4: Reinforce the Roof

Use reinforced concrete, steel, or sandbags to build a strong, blast-resistant roof. Make sure the roof is sloped to deflect debris and prevent water pooling.

If possible, add earth or sandbags on top of the roof for additional radiation shielding.

Step 5: Install Doors and Air Vents

Install a heavy, blast-resistant door made of steel or lead to protect the entrance. Ensure the door seals tightly to prevent fallout from entering.

Place filtered air intake and exhaust vents in strategic positions to maintain airflow and keep fallout particles out. Use HEPA or activated carbon filters to remove contaminants from the air.

Step 6: Add Exterior Shielding

Build an earth berm around the shelter to add additional radiation and blast protection. The berm should be at least 3 feet thick and cover as much of the shelter as possible.

Alternatively, use sandbags, stone, or concrete blocks to reinforce the outside of the shelter.

Step 7: Stock the Shelter

Equip the shelter with enough food, water, and supplies to last at least two weeks. Store canned and dried foods, water in sealed containers, and emergency equipment such as first aid kits, radios, and flashlights.

Include a portable toilet or chemical toilet for sanitation, along with cleaning supplies and waste bags.

Cost-Effective Above-Ground Shelters

If you're on a budget, here are some strategies to reduce the cost of building an above-ground shelter while maintaining safety and functionality:

Use existing structures: Convert a basement, garage, or shed into a fallout shelter by reinforcing the walls and roof with concrete, sandbags, or cinder blocks. Adding external shielding with earth or sandbags can turn an existing building into a viable fallout shelter.

DIY sandbag shelters: Sandbags are an affordable and effective way to create thick, radiation-resistant walls. You can purchase bags and fill them with dirt or sand from your property to keep costs low.

Salvage materials: Use second-hand or salvaged building materials such as cinder blocks, steel beams, or concrete rubble to construct or reinforce the shelter.

Focus on key areas: If you can't afford to reinforce the entire structure with expensive materials like lead or steel, focus on the most critical areas, such as the roof, door, and air vents. Using stronger materials in these areas can still provide a high level of protection.

Above-ground fallout shelters, when built with the right materials and design considerations, can offer significant protection from radiation and blast effects in a nuclear disaster. While they may not provide the same level of shielding as underground shelters, they are more accessible and can be constructed more quickly and affordably. By using dense materials like concrete, sandbags, and cinder blocks, reinforcing the structure to withstand blasts, and ensuring proper ventilation and air filtration, an above-ground shelter can keep you and your family safe during a fallout scenario.

Protecting your family in a fallout zone is a multi-faceted task that requires careful planning, quick decision-making, and long-term preparedness. After a nuclear event, fallout—the radioactive particles that settle to the ground following an explosion—poses one of the greatest dangers. Radiation from fallout can lead to serious health problems, including radiation sickness and cancer, making it essential to know how to protect your family from exposure and ensure long-term survival. In this chapter, we will cover the steps you need to take to safeguard your family in a fallout zone, from emergency actions to long-term sheltering strategies.

Understanding Fallout and Its Dangers

Fallout is composed of tiny radioactive particles that are carried into the atmosphere after a nuclear explosion and eventually settle on the ground. These particles emit harmful radiation, primarily in the form of gamma rays, which can penetrate the skin and damage internal organs. Fallout particles can contaminate the air, water, soil, and surfaces, making exposure unavoidable without proper protection. The highest levels of radiation occur in the first few hours and days following an explosion, but fallout can remain dangerous for weeks or even years, depending on the severity of the event.

The two main dangers from fallout are:

External exposure: Radiation from fallout particles can penetrate the body, causing skin burns, radiation sickness, and long-term health effects such as cancer.

Internal exposure: Ingesting or inhaling radioactive fallout particles can result in internal radiation exposure, which can damage vital organs and increase the risk of cancer.

Immediate Actions After a Nuclear Explosion

In the immediate aftermath of a nuclear explosion, your priority is to minimize radiation exposure and find shelter as quickly as possible. Here are the critical steps to protect your family during the initial stages of a fallout event:

Take Cover Immediately

As soon as you become aware of a nuclear explosion, your first priority is to get your family inside and away from windows and doors. The blast wave from a nuclear explosion can shatter glass and cause severe injuries even miles away from the detonation. If you are outdoors, seek immediate cover in the nearest building or shelter, such as a basement, underground parking structure, or an interior room with no windows.

If you're in a vehicle: Stop and find the nearest sturdy building to take shelter. Vehicles provide very little protection from radiation or blast waves.

If you're outdoors with no shelter nearby: Lie flat on the ground, face down, and cover your head. After the blast wave passes, move to the nearest shelter.

Seal Your Shelter

Once you are inside, take immediate steps to seal the shelter from radioactive fallout particles. Close all windows, doors, and vents, and turn off any HVAC systems to prevent fallout from being drawn inside. If you are in a home or other building, use plastic sheeting, duct tape, towels, or other available materials to cover cracks around windows and doors.

Move to the innermost room or basement: Fallout particles will settle on exterior surfaces, so it's important to move to a room that offers the most shielding from the outside, such as a basement, underground shelter, or an interior room with no windows.

Stay Inside for At Least 24 Hours

The radiation levels from fallout are highest in the first 24 to 48 hours after a nuclear explosion. During this time, you should remain inside the shelter and avoid going outdoors. Radiation levels decrease significantly after the first two days, but staying indoors for as long as possible reduces your overall exposure. In many cases, it's safest to remain in the shelter for at least two weeks if supplies allow.

Monitor Emergency Alerts

Use a battery-powered or hand-crank radio to monitor official emergency broadcasts. Local authorities will provide updates on radiation levels, areas of danger, and when it's safe to evacuate or leave the shelter. Communication systems may be disrupted immediately after the event, so listen for updates regularly.

Protecting Your Family during Extended Shelter Stay

After the immediate threat of the blast has passed, the challenge of long-term survival in a fallout zone begins. Protecting your family during an extended stay in a fallout shelter involves careful management of supplies, sanitation, and psychological well-being.

Water and Food Safety

Your family's survival in a fallout zone depends on having access to clean water and safe food. Contamination of food and water with radioactive fallout is a significant concern, so taking precautions is vital.

Water safety: Ensure you have a sufficient supply of clean, uncontaminated water stored in sealed containers. Do not rely on surface water sources (such as rivers or lakes) unless you have the means to purify them. Use water filters that can remove radioactive particles (such as reverse osmosis systems) or purify water with iodine tablets or boiling.

Food safety: Eat only food that was stored indoors in sealed containers, such as canned goods, dried foods, and freeze-dried meals. Avoid consuming any fresh produce or meat from outside, as it may be contaminated by fallout particles. If you have to use outdoor water or food sources, ensure they are thoroughly tested for radiation first.

Sanitation and Hygiene

Maintaining cleanliness and sanitation in a confined space is essential to prevent illness. With limited water and resources, hygiene may become challenging, but it is still crucial to protect your family from diseases and infections.

Waste management: If your shelter does not have plumbing, use a portable or chemical toilet and dispose of waste in sealed containers or bags. Store waste in a designated area away from the living space until it can be safely removed.

Personal hygiene: Use sanitizing wipes, alcohol-based hand sanitizers, and disposable towels to keep hands and body clean. If water is limited, prioritize drinking water over washing, but clean essential areas like the face, hands, and mouth to reduce the risk of illness.

Clothing: Keep a separate set of clean clothing for each person in the shelter, and change regularly. If any family members must leave the shelter, have them remove and store their contaminated clothing before re-entering to prevent radioactive particles from spreading inside.

Decontaminating After Exposure

If you or a family member has been exposed to radioactive fallout while outside the shelter, immediate decontamination is necessary to reduce the risk of internal radiation exposure.

Remove contaminated clothing: Carefully remove clothing and shoes that may have been exposed to fallout and place them in sealed plastic bags. Contaminated clothing can retain radioactive particles, so dispose of it properly.

Wash exposed skin: Use soap and water to wash exposed skin, including the hands, face, and hair. Pay special attention to areas where fallout may have settled, such as the scalp, under the nails, and on exposed skin. If water is limited, use wet wipes or a damp cloth to remove particles.

Radiation Monitoring

Monitoring radiation levels in your environment is essential for knowing when it is safe to leave the shelter or move outside. Equip your shelter with a Geiger counter or dosimeter, which can measure radiation levels and help you assess the risk of exposure.

Safe radiation levels: Radiation levels are measured in microsieverts per hour (μSv/h) or millisieverts (mSv). Generally, levels below 0.5 μSv/h are considered safe for limited exposure. If levels exceed this, limit outdoor activity and continue sheltering.

Mental Health and Well-Being

Long periods of confinement, fear, and uncertainty can take a toll on your family's mental and emotional well-being. Maintaining a sense of routine, communication, and support is vital to keeping everyone calm and focused during a fallout event.

Create a daily routine: Establish a daily schedule that includes tasks like preparing food, cleaning, monitoring radiation levels, and exercise. Keeping a routine helps maintain a sense of normalcy and reduces stress.

Entertainment and distractions: Make sure your shelter includes items for entertainment and mental stimulation, such as books, puzzles, games, or a radio. These activities help pass the time and keep morale up, especially for children.

Communication and reassurance: Encourage open communication within the family, allowing everyone to express their fears or concerns. Reassure your family by explaining the steps you are taking to keep them safe and discussing plans for the future.

Long-Term Survival in a Fallout Zone

In the worst-case scenario, where radiation levels remain high for an extended period or infrastructure is severely damaged, your family may need to adapt to long-term survival in a fallout zone. Here are some strategies for managing long-term fallout exposure:

Testing and Treating Contaminated Water

Water sources may be contaminated by fallout for months or even years. Continuously monitor radiation levels in local water sources and use filtration methods such as reverse osmosis, distillation, or activated carbon filters to purify water. Rainwater collection systems may also provide a safer water source if collected after the initial fallout has settled.

Growing Food Safely

Growing food in a fallout zone is challenging but possible with the right precautions. Use raised beds filled with uncontaminated soil or greenhouses to isolate plants from radioactive fallout. Select crops that are less likely to absorb radiation, such as fruits or grains, and avoid root vegetables that grow directly in contaminated soil.

Medical Preparedness

Prolonged radiation exposure can lead to health complications, so it's essential to have a well-stocked first aid kit and a supply of any necessary medications. If possible, stock medications like potassium iodide, which can protect the thyroid from absorbing radioactive iodine. Keep in mind that medical assistance may not be readily available in a fallout zone, so be prepared to handle minor injuries and illnesses on your own.

Evacuation Planning

When radiation levels have decreased to safer levels, you may need to plan an evacuation if living in the fallout zone becomes unsustainable. Monitor local radiation levels and official guidance to determine the safest time to leave. Ensure that your family is prepared with a bug-out bag containing essentials like food, water, clothing, and medical supplies.

Surviving in a fallout zone requires careful planning, fast action, and long-term adaptability. By understanding the dangers of radiation exposure and following essential steps to protect your family, you can minimize the risks and increase your chances of survival. From taking immediate shelter to managing food, water, and health in the long term, each step is critical in ensuring that you and your loved ones stay safe in the aftermath of a nuclear event. With the right preparation and mindset, you can face the challenges of a fallout zone and protect your family for the future.

Short-Term Fallout Survival: The First 72 Hours

The first 72 hours after a nuclear explosion are the most critical for survival. During this period, radiation levels are at their highest, and immediate actions are necessary to protect yourself and your family from the deadly effects of radioactive fallout. Short-term survival in a fallout zone requires quick thinking, effective sheltering, and careful resource management to minimize exposure and increase your chances of making it through this dangerous window. In this chapter, we'll cover what you need to do during the first 72 hours after a nuclear event, focusing on sheltering, radiation protection, communication, and basic survival needs.

Why the First 72 Hours Are Critical

In the aftermath of a nuclear explosion, radioactive fallout will begin to settle over the surrounding area within minutes to hours, depending on the size of the blast and weather conditions. Fallout consists of tiny radioactive particles that emit gamma radiation, which can penetrate most materials and cause severe health effects, including radiation sickness and death. Radiation levels will be highest in the first 24 to 48 hours after the explosion, making immediate sheltering and protection essential. While the danger of radiation decreases over time, the first 72 hours are crucial for survival, as this is when the highest radiation exposure occurs.

Immediate Actions: The First Hour

As soon as you are aware of a nuclear explosion, your top priority is to protect yourself and your family from the initial blast and fallout. Taking the right actions in the first hour can make all the difference between life and death.

Take Immediate Shelter

The first priority after a nuclear explosion is to take shelter immediately. Fallout can start falling within minutes, so there is no time to waste. If you are indoors, stay inside and move to the innermost room of your home, away from windows, doors, and exterior walls. If you have access to a basement or underground shelter, go there immediately, as these areas provide the best protection from radiation.

If you're outside: If you are caught outside when a nuclear explosion occurs, seek shelter as quickly as possible. The nearest building, underground structure, or even a culvert or ditch can provide some protection. Cover your mouth and nose with a cloth to avoid inhaling fallout particles, and shield your skin as much as possible. Once inside, do not go outside again unless absolutely necessary.

Seal off the shelter: Close all windows, doors, and vents to prevent radioactive fallout particles from entering your shelter. Use duct tape, plastic sheeting, or towels to seal any cracks or gaps around doors and windows. Stay as far away from exterior walls as possible to minimize radiation exposure.

Duck and Cover

If you are close enough to the explosion to see the flash, you must take immediate steps to protect yourself from the blast wave, which can arrive seconds to minutes after the flash. Use the "duck and cover" technique to minimize injury from flying debris and the shockwave.

Find cover: If you are outdoors, lie face down with your hands covering your head and neck. If you are indoors, take cover under a sturdy table or piece of furniture.

Stay down: The blast wave may arrive within seconds to minutes after the flash, so remain in place until it passes.

The First 24 Hours: Immediate Fallout Protection

Once the immediate danger of the blast has passed, the next critical phase is protecting yourself from radioactive fallout. Fallout particles can begin falling as soon as 15 minutes after the explosion, so remaining in a well-sealed shelter is essential.

Stay Inside

For the first 24 hours after a nuclear explosion, it is crucial that you and your family stay inside your shelter. Fallout particles outside will emit dangerous levels of radiation, and exposure, even for a few minutes, can be deadly. Do not leave your shelter unless absolutely necessary.

Limit exposure: If you must go outside to check on family members or for any other reason, limit your time to no more than a few minutes. Wear a mask or cover your mouth and nose with a cloth to avoid inhaling fallout particles, and remove any contaminated clothing as soon as you return indoors.

Monitor Radiation Levels

If you have access to a Geiger counter or radiation dosimeter, monitor radiation levels regularly to determine the level of exposure in your area. Understanding the level of radiation will help you determine how long you need to remain in your shelter.

Safe levels: Ideally, you want to wait until radiation levels drop to 0.5 microsieverts per hour (μSv/h) or lower before considering any prolonged outdoor activity. However, you should remain indoors for at least the first 24 hours even if radiation levels appear to be dropping quickly.

Decontaminate After Exposure

If you or a family member has been exposed to fallout by going outdoors, it's essential to decontaminate as soon as possible to remove radioactive particles from the skin and hair.

Remove contaminated clothing: Carefully remove clothing and shoes that may have come into contact with fallout and place them in a sealed plastic bag or container. This helps prevent radioactive particles from spreading indoors.

Wash exposed skin: Use soap and water to wash exposed skin thoroughly. Pay close attention to areas where fallout particles may have settled, such as the scalp, under the nails, and the face. If water is limited, use wet wipes or a damp cloth to wipe away particles.

Sustaining Yourself: The First 48 to 72 Hours

Once the immediate threat of the blast and fallout has passed, your focus will shift to surviving inside the shelter. You'll need to manage your resources, ensure everyone's health and well-being, and prepare for the next phase of the emergency.

Water and Food Management

During the first 72 hours, you'll need to carefully manage your water and food supplies to ensure they last until it's safe to leave the shelter or until help arrives. Radiation may have contaminated outside water sources, so avoid using any water that wasn't stored indoors or protected from fallout.

Water storage: Each person needs at least 1 gallon of water per day. Prioritize drinking water over washing or cleaning to conserve supplies. If you need to purify water, use filters or purification tablets.

Food: Focus on eating shelf-stable, non-perishable foods such as canned goods, dried foods, or freeze-dried meals. Avoid fresh food that may have been exposed to fallout particles, and wash any containers or packaging before opening them.

Sanitation and Hygiene

Keeping your shelter clean and maintaining personal hygiene are essential for preventing illness during the first 72 hours. Without access to running water, you'll need to use alternative methods to stay clean and dispose of waste.

Waste disposal: Use a portable toilet or chemical toilet for waste disposal, and store waste in sealed containers or bags. If you don't have a portable toilet, create a designated waste area and use plastic bags for collection. Keep the waste area as far from the living space as possible.

Personal hygiene: Use hand sanitizers, wet wipes, or alcohol-based disinfectants to clean your hands and body. Make sure to clean your hands before eating or handling food. If water is available, wash your face and hands at least once a day to reduce the risk of infection.

First Aid and Health Monitoring

In the first 72 hours, you'll need to be prepared to handle minor injuries, illnesses, and potential radiation sickness. A well-stocked first aid kit is essential for treating cuts, burns, or other injuries caused by the blast or exposure.

Radiation sickness symptoms: Symptoms of radiation sickness include nausea, vomiting, diarrhea, fatigue, and skin burns. If anyone in the shelter starts showing these symptoms, make them as comfortable as possible, keep them hydrated, and monitor their condition closely. Severe cases will require medical attention when it becomes available.

Communication and Information

Staying informed is critical in the first 72 hours after a nuclear explosion. Official updates from local authorities will provide information on radiation levels, evacuation orders, and relief efforts. Use a battery-powered or hand-crank radio to stay updated on the situation.

Check for updates: Tune in to emergency broadcasts regularly for information about the extent of the fallout, areas to avoid, and instructions on when it's safe to evacuate. Keep in mind that communication infrastructure may be disrupted immediately after the event, so it may take time for updates to come through.

Stay connected: If cell phone or internet service is still available, use it sparingly to conserve battery life. Check in with family or emergency services, but avoid unnecessary use to ensure your devices last.

Preparing to Leave: After 72 Hours

After the first 72 hours, radiation levels should begin to decrease, although the danger is far from over. At this point, you'll need to assess the situation carefully and determine whether it's safe to remain sheltered or if you need to prepare for evacuation.

Assess Radiation Levels

Use your Geiger counter or dosimeter to measure radiation levels both inside and outside the shelter. If levels are still high (above 1.0 µSv/h), continue sheltering and limit exposure to the outside. Radiation levels drop significantly after the first 48 to 72 hours, but they may still be dangerous for extended periods in areas close to the blast.

Prepare for Evacuation

If authorities issue an evacuation order or if your shelter is no longer safe, you may need to prepare for evacuation. Gather your emergency supplies, including water, food, first aid, and personal protective equipment like masks or gloves. Ensure that everyone is adequately covered to avoid direct contact with fallout.

Evacuation route: Plan your route based on the latest updates on radiation levels and road conditions. Avoid areas with high radiation readings and head toward designated evacuation centers or safe zones.

Travel precautions: If you must evacuate through a fallout zone, limit your time outdoors as much as possible. Wear protective clothing and cover exposed skin to prevent fallout from settling on your body. Bring water and food, as local supplies may be contaminated.

The first 72 hours after a nuclear explosion are the most dangerous and critical for survival. By taking immediate shelter, sealing your space from fallout, managing your resources, and staying informed, you can protect yourself and your family from radiation exposure. As radiation levels decrease, you'll need to carefully assess the situation and decide whether to continue sheltering or evacuate to a safer area. With the right preparation and quick actions, you can increase your chances of surviving the immediate aftermath of a nuclear event and make it through this dangerous period.

Long-Term Fallout Survival: Months and Years

Long-term survival in a fallout zone involves more than just immediate protection and short-term strategies; it requires careful planning, resource management, and adapting to a radically changed environment for months or even years. After a nuclear event, while the immediate threat from fallout diminishes over time, the effects of radiation and environmental damage can persist for much longer. In this chapter, we'll explore strategies for long-term survival, including securing sustainable food and water supplies, managing health risks, rebuilding infrastructure, and ensuring mental and emotional well-being in a post-fallout world.

Understanding Long-Term Fallout Risks

As time passes after a nuclear event, radiation levels will gradually decrease, but they may not return to safe levels for months or even years in areas that received heavy fallout. The half-life of certain radioactive isotopes, such as cesium-137 and strontium-90, means that they can persist in the environment for decades, continuing to pose health risks through contaminated soil, water, and food sources.

While the immediate risk of acute radiation sickness decreases after the first few weeks, long-term exposure to lower levels of radiation can lead to chronic health issues, including an increased risk of cancer and other illnesses. Understanding these risks and taking steps to minimize exposure over the long term is essential for survival.

Long-Term Food Security: Growing and Sourcing Safe Food

One of the greatest challenges in a post-nuclear environment is ensuring a reliable and safe food supply. Fallout can contaminate soil, plants, and animals, making it dangerous to consume anything grown or harvested in areas affected by radiation. However, with proper planning and precautions, it's possible to grow and source food safely.

Growing Food Safely

To grow food safely in a fallout zone, you must avoid contamination from radioactive particles that may have settled on the soil. This requires using uncontaminated soil, growing in protected environments, and carefully selecting crops.

Raised beds and greenhouses: Raised garden beds filled with fresh, uncontaminated soil are ideal for growing food in a fallout environment. Greenhouses provide additional protection by shielding crops from fallout particles and allowing you to control the growing environment.

Crops to grow: Certain crops are less likely to absorb radiation from the soil. Focus on growing fruits and grains, which tend to accumulate fewer radioactive isotopes than leafy greens or root vegetables. Avoid crops like carrots, potatoes, and turnips, which grow directly in the soil and are more prone to contamination.

Testing soil: Regularly test the soil for radiation levels using a Geiger counter or other radiation detection devices. If radiation levels are too high, refrain from growing food in that soil until it becomes safe.

Hunting, Fishing, and Foraging

Hunting, fishing, and foraging may become necessary for long-term survival, but you must be cautious of contamination in wild animals, fish, and plants.

Hunting and fishing: Animals and fish that have ingested radioactive particles can pose a significant health risk if consumed. Test any meat or fish with a radiation detector before eating, and avoid consuming any that show signs of contamination. Consider sticking to animals and fish from areas known to have lower fallout exposure.

Foraging: While foraging for wild plants and berries may seem like a good food source, these plants can also absorb radiation from contaminated soil. Only forage from areas that have been tested for radiation, and avoid consuming plants that grow close to the ground or in areas of known fallout.

Preserving and Stockpiling Food

In the long-term, preserving and stockpiling food is essential to ensure a steady supply of uncontaminated food. Proper preservation methods allow you to store food safely for months or years, reducing reliance on potentially contaminated sources.

Canning: Home canning is one of the best ways to preserve food for long-term storage. Fruits, vegetables, and meats can be safely canned and stored in sealed jars, keeping them free from contamination.

Dehydrating: Dehydrating fruits, vegetables, and meats removes moisture, making them less prone to spoilage. Dehydrated food can be stored for long periods without refrigeration and rehydrated when needed.

Freeze-drying: Freeze-drying is another excellent preservation method, as it retains the nutritional value of the food while extending its shelf life. Although freeze-drying equipment can be expensive, it provides a reliable source of long-term food.

Securing a Long-Term Water Supply

Water is crucial for long-term survival, and ensuring a safe, uncontaminated water source is one of the most difficult challenges in a fallout environment. Water sources like rivers, lakes, and even groundwater can be contaminated by radioactive particles, so finding and purifying water is essential.

Water Purification

Radioactive particles in water can cause long-term health problems if ingested, so purifying any water you collect is essential for survival.

Filtration systems: Use advanced water filtration systems, such as reverse osmosis filters, to remove radioactive particles from contaminated water. These filters are capable of removing the smallest particles, including radioactive isotopes.

Distillation: Distillation is one of the most effective ways to purify water in a fallout zone. By boiling water and collecting the condensed vapor, you can remove radioactive particles and other contaminants. While distillation requires significant energy, it provides a safe and reliable source of clean water.

Rainwater collection: Rainwater collected after the initial fallout period can be a safer alternative to contaminated surface water. Use a rainwater collection system with proper filtration to store clean water for long-term use.

Water Storage

Storing large quantities of water is essential for long-term survival, as access to clean water may be inconsistent. Water should be stored in sealed, durable containers to protect it from contamination.

Plastic barrels or tanks: Store water in food-grade plastic barrels or tanks that can hold large volumes of water. Ensure the containers are sealed tightly to prevent fallout from entering.

Rotating stored water: Even stored water should be rotated regularly to prevent stagnation or contamination. Make a habit of using and replacing stored water periodically to ensure it remains fresh.

Health and Medical Care in a Fallout Zone

Long-term survival in a fallout zone requires careful attention to health and medical care. Exposure to radiation, contaminated food, and water can lead to chronic health problems, so taking steps to protect your family's health is essential.

Managing Radiation Exposure

While radiation levels will decrease over time, long-term exposure to even low levels of radiation can cause serious health issues, including cancer. Managing radiation exposure involves reducing contact with contaminated areas, food, and water.

Radiation monitoring: Use a Geiger counter or dosimeter to regularly monitor radiation levels in your environment. If levels are too high, avoid spending extended time outdoors, especially in areas of heavy fallout.

Potassium iodide: Keep a supply of potassium iodide (KI) on hand. KI protects the thyroid gland from absorbing radioactive iodine, reducing the risk of thyroid cancer. It's most effective when taken shortly before or after exposure to radioactive iodine.

Maintaining Medical Supplies

In a long-term survival situation, access to medical care may be limited or non-existent. Having a well-stocked first aid kit and essential medications is critical for addressing injuries and illnesses.

Stockpile medications: If possible, stockpile prescription medications that your family depends on, as access to pharmacies or medical facilities may be limited. Include over-the-counter medications such as pain relievers, antacids, and allergy medications in your stockpile.

First aid supplies: A comprehensive first aid kit should include bandages, antiseptics, sterile gauze, scissors, splints, and other essentials for treating injuries. Learn basic first aid and CPR to handle emergencies.

Treating radiation sickness: Radiation sickness can develop over time with prolonged exposure. Symptoms include nausea, vomiting, fatigue, and skin burns. While severe cases require professional medical treatment, keeping the affected person hydrated and rested can help manage milder symptoms.

Rebuilding Infrastructure and Community

In a post-nuclear world, the infrastructure you once relied on may be destroyed or severely compromised. Rebuilding a sustainable, self-sufficient way of life will be essential for long-term survival. Re-establishing community ties and working together with others will increase your chances of rebuilding and thriving.

Shelter and Home Maintenance

After months or years in a fallout zone, your shelter or home will require maintenance or rebuilding to ensure it remains safe and habitable. Radiation may have weakened materials, and exposure to the elements can cause wear and tear on the structure.

Reinforce your shelter: As radiation levels decrease, assess the condition of your shelter and reinforce any areas that have weakened or been damaged. Repair cracks in walls, reseal doors and windows, and reinforce the roof if necessary.

Expand living spaces: If conditions improve, you may be able to expand your shelter to accommodate more people or activities, such as food storage or indoor gardening.

Community and Cooperation

Working with others is essential for long-term survival in a post-fallout world. Building or joining a community of like-minded individuals allows you to share resources, knowledge, and skills, making survival easier and more sustainable.

Form alliances: Establish relationships with other survivors in your area and work together to create a community. Pooling resources and skills can help everyone thrive in the long term.

Bartering and trade: In the absence of a functioning economy, bartering will become a vital way to acquire necessary goods. Stockpile items that are valuable for trade, such as food, water, medical supplies, and tools.

Mental and Emotional Well-Being

Surviving in a fallout zone for months or years takes a toll on your mental and emotional health. Prolonged isolation, stress, and fear can lead to depression, anxiety, and other psychological issues. Taking care of your mental health is just as important as ensuring your physical survival.

Maintaining a Routine

Establishing a daily routine helps maintain a sense of normalcy and control, reducing stress and anxiety. Include regular tasks such as preparing meals, cleaning, tending to plants or animals, and monitoring radiation levels in your routine.

Staying Occupied

Keeping your mind occupied can help prevent feelings of boredom, depression, or hopelessness. Engage in activities such as reading, writing, crafting, or learning new skills to pass the time and stay mentally active.

Social Interaction

If you are part of a community, maintaining social interactions is crucial for emotional well-being. Regular communication, shared activities, and mutual support can boost morale and create a sense of purpose. Even within a small family unit, talking and sharing feelings helps alleviate stress.

Mental Health after the Blast: Coping with Trauma

Surviving a nuclear event is not just a physical challenge—it's an emotional and psychological one as well. The trauma of witnessing such an extreme disaster, combined with the stress of long-term survival in a fallout zone, can take a profound toll on mental health. In the aftermath of a nuclear blast, fear, uncertainty, loss, and isolation become constant companions. Coping with trauma and maintaining mental health will be critical for both individual well-being and the ability to make sound decisions during survival. In this chapter, we will explore how to manage the psychological effects of a nuclear disaster, strategies for coping with trauma, and steps to foster mental resilience in yourself and your family.

Understanding the Psychological Impact of a Nuclear Event

The psychological effects of surviving a nuclear disaster can be overwhelming. The sudden and catastrophic nature of the event can lead to a range of emotional responses, including shock, fear, anger, grief, and guilt. The long-term challenges of living in a fallout zone—scarcity of resources, loss of social connections, isolation, and the constant threat of radiation—can further exacerbate stress, anxiety, and depression.

Here are some common psychological reactions that survivors may experience:

Acute stress reactions: In the immediate aftermath of the disaster, survivors often experience heightened anxiety, confusion, and difficulty concentrating. These acute stress reactions are normal, but if they persist, they can evolve into more serious mental health conditions.

Post-traumatic stress disorder (PTSD): PTSD can develop after experiencing or witnessing a traumatic event, such as a nuclear explosion. Symptoms may include flashbacks, nightmares, hypervigilance, and avoidance of reminders of the event.

Survivor's guilt: Survivors may feel guilty about having lived through the disaster when others did not. This guilt can manifest in feelings of unworthiness or shame, and survivors may question their decisions during the event.

Depression and hopelessness: The long-term survival challenges, combined with isolation and the destruction of normal life, can lead to depression. Feelings of hopelessness, sadness, and loss of interest in life are common in survivors.

Anxiety and fear: Ongoing anxiety about radiation exposure, lack of resources, and the uncertainty of the future can weigh heavily on survivors' mental well-being. Chronic fear can lead to physical symptoms like headaches, fatigue, and insomnia.

Coping with Trauma: Short-Term Strategies

In the immediate aftermath of the nuclear event, your first priority will be to focus on survival and physical safety. However, addressing the emotional and psychological impact is equally important. Here are some strategies for coping with trauma in the short term:

Acknowledge and Process Emotions

After a traumatic event, it's normal to experience a range of intense emotions, from fear and anger to sadness and numbness. Instead of suppressing these emotions, it's important to acknowledge and process them. Give yourself and your family permission to feel upset, anxious, or scared—it's a natural response to a disaster.

Talk about your feelings: If you are sheltering with family or others, encourage open conversations about how everyone is feeling. Talking through emotions can help relieve some of the mental pressure and allow for mutual support.

Journaling: If you find it difficult to talk about your feelings, consider writing them down. Journaling can be a powerful way to process emotions and reflect on the situation, helping you gain a sense of control.

Establish a Routine

In the chaos following a nuclear event, creating a daily routine can provide a sense of stability and control, which helps reduce anxiety and stress. A structured routine helps anchor your day, offering a sense of normalcy amidst the uncertainty.

Daily tasks: Set regular times for tasks such as preparing meals, cleaning, and checking supplies. Assign responsibilities to each family member to ensure that everyone stays engaged and focused.

Include breaks and relaxation: While it's important to stay busy, it's equally important to take breaks. Incorporate short periods of relaxation or quiet time into your routine, such as meditation, deep breathing exercises, or simply sitting quietly to clear your mind.

Stay Connected with Loved Ones

Social connection is a powerful tool for coping with trauma. In the aftermath of a nuclear disaster, your family and shelter companions will be your primary source of emotional support. Encourage communication and cooperation, and lean on each other to navigate the challenges ahead.

Support each other: Encourage family members to share their thoughts and worries. Provide emotional support by listening without judgment and offering reassurance. Knowing that you are not alone in facing these difficulties can reduce feelings of isolation.

Use technology (if available): If phone or internet access is still functioning, try to stay in contact with extended family, friends, or even online communities of survivors. Even brief check-ins can provide comfort and reduce the feeling of being cut off from the outside world.

Focus on Immediate Survival Goals

In a crisis, focusing on short-term goals can help reduce overwhelming feelings of helplessness. Rather than getting lost in fear of the future, concentrate on what you can control in the present.

Break tasks into steps: Focus on immediate survival needs, such as securing food, water, and shelter. Break larger tasks into smaller, manageable steps. For example, instead of worrying about long-term food shortages, focus on rationing supplies for the next few days.

Celebrate small wins: Completing tasks, even small ones, can give you a sense of accomplishment and control. Whether it's reinforcing the shelter or purifying water, acknowledge the progress you are making toward survival.

Long-Term Coping: Months and Years After the Blast

As time passes and the initial shock wears off, the long-term psychological effects of surviving in a fallout zone may begin to emerge. Loneliness, despair, and the weight of prolonged survival can become heavy burdens. Developing long-term coping mechanisms is crucial to maintaining mental health and resilience in the months and years that follow.

Set Long-Term Survival Goals

After the initial crisis period, it's important to shift focus to long-term survival strategies. Setting realistic, achievable goals for rebuilding and sustaining life can provide a sense of purpose and direction, which helps mitigate feelings of hopelessness.

Rebuilding efforts: Identify areas of your life that need rebuilding, whether it's improving the shelter, securing a more reliable food source, or creating a community with other survivors. Set long-term goals, such as growing a sustainable garden or establishing trade with others, to give yourself and your family something to work toward.

Adapt to the new normal: Accepting that life has changed after the blast is essential for moving forward. While mourning the loss of the life you once knew is important, focusing on adapting to the new reality will help you find peace in the present.

Manage Depression and Anxiety

Long-term survival can lead to chronic stress, depression, and anxiety. If left unchecked, these feelings can make it difficult to function and maintain hope for the future. Recognizing the signs of depression and anxiety and addressing them early is critical.

Exercise: Physical activity is a proven method for improving mood and reducing stress. In a confined space, exercises like stretching, yoga, or bodyweight exercises can release tension and improve mental well-being.

Mindfulness and meditation: Practicing mindfulness or meditation can help you stay present and focused, reducing anxiety about the past or future. Spend a few minutes each day focusing on your breath, calming your mind, and letting go of worries.

Seek meaning: Finding purpose or meaning in your survival can help combat feelings of hopelessness. This might involve taking care of your family, working toward long-term goals, or even helping others in your community.

Combat Isolation and Loneliness

The isolation that comes with living in a fallout shelter or cut off from the outside world can lead to intense loneliness. Over time, this isolation can have serious mental health effects, such as depression and feelings of abandonment.

Create a sense of community: If you are part of a small group or family, cultivate a strong sense of togetherness and community. Regularly check in with each other, engage in group activities, and share responsibilities. Mutual support and cooperation are essential for combating loneliness.

Form connections with other survivors: If it's safe to do so, establish contact with other survivors in your area. This could be done through radio communication or meeting in person once radiation levels decrease. Forming alliances and rebuilding social ties can provide a crucial sense of connection.

Help Children Cope with Trauma

Children are particularly vulnerable to the psychological effects of trauma, and they may have difficulty understanding the scope of the disaster or expressing their emotions. Helping children cope with the aftermath of a nuclear event requires patience, understanding, and reassurance.

Maintain routines: Children find comfort in routine, so maintaining a regular schedule of meals, activities, and sleep will help them feel safe and secure.

Talk openly about feelings: Encourage children to talk about their feelings and fears. Answer their questions as honestly as possible while offering reassurance. Use age-appropriate language to explain what has happened and what you're doing to keep them safe.

Engage in play: Play is a natural way for children to process emotions and trauma. Encourage them to play games, draw, or use their imagination to work through their feelings in a safe and healthy way.

Building Mental Resilience

Long-term survival in a post-nuclear world requires mental resilience—the ability to adapt to adversity, recover from trauma, and maintain hope. While it may feel impossible to stay positive in such a challenging situation, there are ways to build resilience and improve your psychological well-being.

Stay Focused on What You Can Control

In a world filled with uncertainty and danger, focusing on what you can control helps reduce feelings of helplessness. Concentrate on the actions you can take to improve your situation, no matter how small.

Practice Gratitude

Even in difficult circumstances, practicing gratitude can have a powerful effect on your mental outlook. Each day, take a moment to reflect on something you're thankful for—whether it's your family's safety, a successful task, or simply having shelter. Gratitude helps shift your mindset from fear to hope.

Maintain a Sense of Humor

Laughter is a natural stress reliever and can lighten the emotional weight of survival. Even in dark times, finding moments of humor—whether through telling stories, playing games, or reminiscing—can help lift spirits and bring people together.

Have Hope for the Future

Maintaining hope is one of the most important aspects of mental resilience. Even when the situation seems dire, focusing on the possibility of recovery, rebuilding, and the future can help keep you going. Remember that human beings have survived and thrived through countless challenges, and with persistence, you can too.

Protecting Livestock and Pets in a Fallout Area

When preparing for survival in a fallout zone, it's not just humans who need protection—your livestock and pets also face significant dangers from radiation, contaminated food and water, and exposure to fallout particles. These animals are vital not only for companionship but also for long-term survival, especially livestock that may provide food and resources in a post-disaster world. In this chapter, we'll explore strategies for protecting livestock and pets in a fallout area, including sheltering, feeding, water management, and long-term care.

The Risks to Animals in a Fallout Area

Like humans, animals can suffer from radiation sickness, internal and external contamination, and long-term health effects if exposed to radioactive fallout. Protecting them from fallout involves minimizing their exposure to contaminated air, water, and food while providing a safe and secure environment.

The primary risks to livestock and pets in a fallout zone are:

Radiation exposure: Direct exposure to gamma radiation from fallout can cause radiation sickness in animals, leading to symptoms such as vomiting, diarrhea, hair loss, and, in severe cases, death.

Ingestion of radioactive particles: Animals that ingest fallout-contaminated food or water are at risk of internal radiation exposure, which can damage organs and increase the risk of cancers.

Inhalation of radioactive dust: Breathing in radioactive dust or particles can cause lung damage and increase the risk of radiation sickness.

Protecting Livestock from Fallout

Livestock—such as cows, chickens, pigs, sheep, and goats—play a critical role in long-term survival, providing meat, milk, eggs, and other essential resources. Protecting them from fallout is crucial not only for their well-being but also for ensuring the sustainability of your food supply.

Sheltering Livestock

The first step in protecting livestock from fallout is providing adequate shelter. Just like humans, livestock need to be shielded from direct exposure to radioactive particles and radiation.

Move livestock indoors: If possible, bring all livestock into barns, sheds, or other indoor shelters. These structures will offer protection from fallout particles, reducing radiation exposure. Ensure that the shelter is sealed as much as possible to prevent fallout dust from entering.

Build makeshift shelters: If you don't have existing barns or sheds, consider constructing makeshift shelters using available materials like tarps, plywood, or plastic sheeting. While these shelters may not be as robust as permanent structures, they can still offer some protection from fallout.

Reinforce shelter walls: For maximum radiation protection, reinforce the walls of livestock shelters with dense materials like sandbags, earth, or cinder blocks. These materials can help block gamma radiation, similar to how they protect human fallout shelters.

Limit outdoor exposure: In the early days after a nuclear event, radiation levels will be at their highest. Keep livestock indoors for at least the first 48 to 72 hours when fallout is most dangerous. After this period, limit their time outside until radiation levels decrease to safer levels.

Providing Clean Food and Water

One of the biggest challenges in caring for livestock after a nuclear disaster is ensuring they have access to uncontaminated food and water. Radioactive fallout can contaminate grass, hay, feed, and water sources, posing a severe risk to livestock health.

Stockpile feed: Before a nuclear event, stockpile enough clean, uncontaminated feed to last at least two weeks. Store feed in airtight containers or sealed bags to prevent fallout contamination. If you are caught unprepared, cover existing feed with tarps or plastic sheeting to protect it from fallout dust.

Keep water protected: Ensure that livestock have access to clean, uncontaminated water. If possible, store water in large containers or barrels inside the shelter. Cover outdoor water troughs or buckets with lids or tarps to prevent fallout particles from entering.

Water filtration: If your livestock water supply has been contaminated, use a filtration system to remove radioactive particles before allowing the animals to drink. You can also purify water through methods like distillation, which effectively removes contaminants.

Monitoring Livestock Health

Radiation exposure can lead to health issues in livestock that are not immediately apparent. It's essential to monitor your animals for signs of radiation sickness and other health problems in the days and weeks following a nuclear event.

Symptoms of radiation sickness: Watch for signs such as lethargy, loss of appetite, diarrhea, vomiting, excessive shedding, skin burns, or open sores. If you observe these symptoms, isolate the affected animal to prevent further contamination and provide supportive care (hydration, rest, and clean shelter).

Regular health checks: Perform daily health checks on your livestock, inspecting them for signs of illness or injury. Keeping a record of their condition will help you identify any changes that could indicate radiation exposure or other health problems.

Long-Term Livestock Care

As radiation levels decrease over time, it's important to ensure your livestock have the resources they need to survive in the long term. This includes securing safe grazing areas, re-establishing a clean water supply, and protecting them from secondary hazards, such as predators or looters.

Safe grazing areas: After the initial fallout period has passed, begin testing grazing areas for radiation before allowing your animals to roam. Use a Geiger counter to measure radiation levels in the grass and soil. Only allow grazing in areas where radiation levels have decreased to safe levels (typically below 0.5 µSv/h).

Rotate pastures: To minimize radiation exposure, rotate your livestock between different pastures. This helps prevent prolonged exposure to contaminated soil and allows time for radiation levels in specific areas to decrease.

Secure your livestock: In a post-nuclear environment, social disorder and scarcity of resources may lead to increased theft or predation. Ensure your livestock are securely sheltered at night, and consider reinforcing fences or barriers to protect them from potential threats.

Protecting Pets from Fallout

Pets, including dogs, cats, and other domestic animals, are beloved members of the family, and their safety should be a priority in a fallout situation. Pets are vulnerable to the same dangers as humans—radiation exposure, contaminated food and water, and inhalation of radioactive particles—so it's crucial to take steps to protect them.

Sheltering Pets Indoors

The best way to protect pets from fallout is to keep them indoors, just as you would for yourself and your family. If your home has a designated fallout shelter, bring your pets into the shelter with you.

Limit outdoor exposure: Keep pets inside the shelter or home for at least the first 48 to 72 hours after a nuclear event, as this is when fallout will be most dangerous. After this period, limit their time outdoors, and ensure they are supervised when they do go outside.

Provide a clean area: Designate a specific area in the shelter for your pets, including bedding, toys, and food. Ensure the area is free of fallout contamination by regularly cleaning it and keeping it sealed from the outside environment.

Providing Safe Food and Water for Pets

Like livestock, pets need access to clean, uncontaminated food and water during a fallout event. Stockpiling enough food and water for your pets is essential to ensure their survival.

Stockpile pet food: Keep a supply of dry or canned pet food stored in airtight containers. Make sure the food is stored in a safe, sealed location to prevent fallout contamination. Plan for at least two weeks of food for each pet, and ration it carefully to ensure supplies last.

Protect water sources: Provide your pets with clean, fresh water that has been stored indoors or in sealed containers. Avoid giving them water from outdoor sources, such as puddles, ponds, or lakes, as these may be contaminated with fallout.

Water purification: If your pet's water supply becomes contaminated, use a water purification method, such as filtration or distillation, to ensure it is safe for them to drink.

Decontaminating Pets

If your pets go outside during or after a fallout event, they may be exposed to radioactive particles. It's essential to decontaminate them as soon as they come back indoors to prevent radiation exposure and contamination of your shelter.

Wash exposed fur: If your pet has been outside, thoroughly wash their fur with soap and water to remove any fallout particles. Pay special attention to their paws, ears, and face, as these areas are most likely to come into contact with fallout.

Remove contaminated clothing or collars: If your pet wears a collar or clothing that has been exposed to fallout, remove and dispose of these items to prevent contamination from spreading indoors.

Monitoring Pet Health

Pets can suffer from radiation sickness just like humans, so it's important to monitor their health closely for signs of illness or radiation exposure.

Symptoms of radiation sickness in pets: Watch for symptoms such as vomiting, diarrhea, excessive drooling, loss of appetite, fatigue, and hair loss. If your pet shows these symptoms, provide them with a quiet, clean space to rest and ensure they have access to fresh water. Seek veterinary care as soon as it becomes available.

Regular check-ups: Perform regular health checks on your pets to ensure they are not suffering from any hidden illnesses or injuries. In the long term, monitor for any signs of radiation-induced illnesses, such as cancers or organ damage.

Long-Term Pet and Livestock Care

Once radiation levels have decreased to safer levels, you'll need to adjust your long-term care strategies for your pets and livestock. Maintaining their health and well-being is essential for long-term survival and companionship.

Testing the Environment

Before allowing pets or livestock to roam outdoors freely, test the environment for radiation levels. Use a Geiger counter to measure radiation in the air, soil, and vegetation, ensuring that levels are safe before giving your animals access to outdoor spaces.

Rebuilding a Food and Water Supply

As radiation levels decrease, you'll need to establish a more sustainable food and water supply for your animals. For livestock, this may mean growing feed or establishing safe grazing areas. For pets, it could involve hunting or foraging for additional food sources.

Growing safe feed: For livestock, begin growing crops for feed in raised beds or greenhouses filled with uncontaminated soil. Rotate pastures to ensure they remain safe for grazing.

Hunting and foraging for pets: In a long-term survival scenario, consider supplementing your pets' diet with wild game or foraged food. Ensure that any food you provide them is tested for contamination before consumption.

Protecting your livestock and pets in a fallout zone requires careful planning, immediate action, and long-term care. By providing safe shelter, uncontaminated food and water, and monitoring their health, you can ensure the survival of your animals through the most dangerous periods of fallout. In the long-term, maintaining a sustainable food and water supply and rebuilding secure environments for your animals will be key to their continued well-being. With the right strategies, you can keep your beloved pets and valuable livestock safe in a post-nuclear world.

Growing Vegetables in Polluted Soil: A Guide to Safe Gardening

In a post-nuclear world, growing your own vegetables is vital for long-term survival, but doing so in soil contaminated by radioactive fallout presents significant challenges. Polluted soil can absorb harmful radioactive isotopes, making it unsafe for growing food without careful precautions. However, with the right strategies and practices, it is possible to cultivate vegetables in such conditions while minimizing health risks. In this chapter, we'll explore how to grow vegetables safely in polluted soil, covering soil preparation, selecting crops, testing for contamination, and long-term gardening techniques.

The Dangers of Growing Vegetables in Polluted Soil

Radioactive fallout can contaminate soil with harmful isotopes such as cesium-137, strontium-90, and iodine-131, which are absorbed by plants and can be ingested when you eat the vegetables. These radioactive elements are especially dangerous because they emit gamma radiation, which can cause radiation sickness, increase the risk of cancer, and lead to long-term health complications.

The main dangers of growing vegetables in contaminated soil include:

Absorption of radioactive particles: Plants grown in contaminated soil can absorb radioactive isotopes through their roots. These isotopes accumulate in plant tissues and can enter the human body when the vegetables are consumed.

Surface contamination: Fallout particles can also settle on the surface of plants, contaminating leaves, stems, and fruits.

Given these risks, it's crucial to use proper techniques to minimize the absorption of radioactive particles and reduce exposure when growing food in a fallout-affected environment.

Strategies for Safe Gardening in Polluted Soil

Despite the challenges, you can still grow vegetables safely by using specific techniques that limit the exposure of plants to radioactive particles. Here's how to prepare and manage your garden in a fallout zone.

Raised Garden Beds and Containers

One of the most effective ways to avoid contaminated soil is to grow vegetables in raised garden beds or containers filled with clean, uncontaminated soil. This keeps your crops from directly contacting polluted ground, minimizing the risk of radiation absorption.

Raised beds: Build raised garden beds using materials like wood, bricks, or cinder blocks, and fill them with fresh soil that has not been exposed to fallout. Raised beds also allow for better control of drainage and soil quality.

Containers: If space or materials for raised beds are limited, use containers such as large pots, barrels, or even buckets to grow your vegetables. Containers can be moved to safer locations and are easier to manage in a contaminated environment.

Soil sourcing: If you don't have access to uncontaminated soil, consider using soil from indoor sources, greenhouses, or soil that was stored in sealed bags before the fallout. If new soil is unavailable, you may need to purify or replace contaminated soil (more on this below).

Soil Remediation: Reducing Contamination

If you must grow vegetables directly in contaminated soil, you can use soil remediation techniques to reduce radiation levels and make the soil safer for gardening. These methods aim to remove or reduce the concentration of radioactive particles in the soil.

Phytoremediation: Phytoremediation is the process of using certain plants to absorb radioactive isotopes from the soil. Plants such as sunflowers, mustard, and Indian grass are known for their ability to absorb radiation. These plants can be grown in contaminated soil and then removed and safely disposed of once they have absorbed the radioactive particles. After several cycles, the soil's radiation levels can decrease.

Soil amendments: Adding materials such as clay, activated charcoal, or zeolite to the soil can help bind radioactive particles, reducing their bioavailability to plants. This prevents plants from absorbing the harmful isotopes. Organic matter such as compost can also help dilute the concentration of radioactive particles in the soil.

Removing the topsoil: If possible, remove the top 6 to 12 inches of contaminated soil and replace it with fresh, uncontaminated soil. Most radioactive particles settle on the surface of the soil, so removing this layer can significantly reduce contamination.

Testing Soil for Contamination

Before planting any vegetables, it's essential to test the soil for radiation levels. This will give you a clear idea of how much contamination is present and whether it's safe to grow crops.

Geiger counter: Use a Geiger counter to test the radiation levels in the soil. While Geiger counters measure overall radiation, they may not provide specific information about the types of radioactive isotopes present. However, if radiation levels are dangerously high, you should avoid using that soil for gardening.

Soil testing kits: Specialized soil testing kits can detect specific radioactive isotopes, such as cesium-137 or strontium-90, in the soil. These kits are more expensive but provide more detailed information about contamination levels. Use the results to decide whether remediation is needed or if you should avoid planting in that area altogether.

Choosing the Right Crops

Certain vegetables are more likely to absorb radioactive particles from the soil than others. When growing food in potentially contaminated areas, it's important to choose crops that are less likely to accumulate radiation. Some plants absorb fewer radioactive particles, making them safer to eat.

Low-absorption crops: Fruits, grains, and vegetables that grow above the ground tend to absorb less radiation than root vegetables. Some safe crops to consider include:

Tomatoes

Peppers

Beans

Squash

Cucumbers

Corn

High-absorption crops: Root vegetables like carrots, potatoes, beets, and turnips are more likely to absorb radiation from contaminated soil, as they grow directly in the ground. If possible, avoid growing these crops in areas with known contamination.

Fast-growing crops: Choose fast-growing crops that can be harvested quickly, reducing the amount of time they are exposed to radiation in the soil. Leafy greens like spinach and lettuce, as well as herbs like basil and cilantro, grow quickly and can be harvested within a few weeks.

Protecting Plants from Fallout

Even if the soil is uncontaminated, fallout particles can settle on the surface of your plants, posing a risk when you harvest and eat the vegetables. To protect your crops, take steps to shield them from exposure to airborne fallout.

Use row covers: Cover your garden with protective row covers made of plastic or other materials to shield plants from fallout particles. These covers should be installed over hoops or stakes to keep them off the plants, allowing airflow while protecting them from contamination.

Greenhouses: If possible, grow vegetables in a greenhouse to provide a controlled environment that keeps fallout particles out. Greenhouses also allow you to control temperature, humidity, and light, ensuring better growing conditions.

Washing crops: Before consuming any vegetables, thoroughly wash them to remove surface contamination. Use clean water, and scrub the vegetables gently to ensure that no fallout particles remain. Peeling vegetables can also reduce the risk of consuming radioactive particles.

Long-Term Gardening in a Fallout Zone

Over time, as radiation levels decrease, you can begin to expand your gardening efforts and grow a wider variety of crops. However, continued vigilance is required to ensure the safety of your garden and food supply.

Rotating Crops

Crop rotation is a valuable practice in any garden, but it becomes especially important in a fallout zone. By rotating crops, you reduce the risk of depleting soil nutrients and prevent plants from repeatedly absorbing any remaining radioactive particles.

Alternating crops: Rotate high-absorption crops (such as root vegetables) with low-absorption crops (such as grains or fruits). This practice allows the soil to recover and reduces the buildup of radioactive particles in the plants.

Cover crops: Plant cover crops like clover or alfalfa to improve soil health, prevent erosion, and help reduce contamination between growing seasons. Cover crops can also contribute organic matter, improving soil structure and nutrient availability.

Long-Term Soil Health

Maintaining soil health is essential for the success of any garden, but in a fallout zone, it becomes even more critical. Focus on improving the soil's organic matter and ensuring proper drainage to minimize contamination risks.

Composting: Create a compost system to recycle organic waste and improve soil fertility. Composting adds vital nutrients to the soil and enhances its ability to support healthy plant growth. However, make sure the materials you use for composting are free from contamination.

Mulching: Use mulch to protect the soil from further fallout contamination and to help retain moisture. Organic mulch, such as straw, leaves, or wood chips, also adds nutrients to the soil as it decomposes.

Soil testing: Continue to test your soil regularly to monitor radiation levels over time. As radiation decreases, you may be able to expand your gardening efforts and grow a wider variety of crops.

Sustainable Water Management

Water is essential for gardening, but in a fallout zone, finding a clean and reliable water source can be difficult. You'll need to ensure that the water you use for irrigation is free from radioactive contamination.

Rainwater harvesting: Rainwater can be a safer alternative to surface water sources that may be contaminated with fallout. Install a rainwater collection system with proper filtration to provide clean water for your garden.

Water filtration: If you must use water from a potentially contaminated source, filter it before using it in your garden. Reverse osmosis systems and activated carbon filters can help remove radioactive particles from the water, making it safer for irrigation.

Growing vegetables in polluted soil after a nuclear event is a challenge, but with the right precautions and strategies, it is possible to produce safe, healthy food. By using raised beds, testing soil for contamination, choosing low-absorption crops, and implementing soil remediation techniques, you can create a sustainable garden even in a fallout zone. Long-term gardening success requires ongoing attention to soil health, water quality, and plant protection, but with perseverance, you can ensure a steady food supply for the future.

Hydroponics for Radiation-Free Food

Hydroponics offers a practical and efficient solution for growing radiation-free food in a fallout zone or any environment where soil contamination is a concern. Since hydroponics doesn't rely on soil, it allows you to cultivate plants in a controlled, enclosed environment where they are shielded from radioactive particles and other contaminants. This chapter will explore the benefits of hydroponics for growing food in a post-nuclear world, the basic components of a hydroponic system, and how to set up and maintain one for long-term, radiation-free food production.

Why Hydroponics is Ideal for a Fallout Environment

Hydroponics is a method of growing plants without soil, using a nutrient-rich water solution to deliver the essential minerals directly to the plant roots. This controlled environment not only eliminates the risk of soil contamination but also allows for more efficient use of water and space, making it ideal for growing food in enclosed shelters or fallout zones.

Here are some reasons why hydroponics is especially suited for growing food in a post-nuclear environment:

No soil contamination: By eliminating the need for soil, hydroponics ensures that plants are not exposed to radioactive particles, heavy metals, or other pollutants commonly found in fallout-affected areas.

Controlled environment: Hydroponic systems are typically set up indoors or in greenhouses, allowing you to control temperature, humidity, and light levels. This isolation from the outside world keeps your plants safe from airborne radioactive particles or other environmental toxins.

Efficient use of water: Hydroponic systems use significantly less water than traditional soil-based gardening, making it a great option in environments where clean water is scarce. The water in a hydroponic system can be recirculated and reused, minimizing waste.

Faster plant growth: Hydroponically grown plants tend to grow faster than those grown in soil because they receive a precise mix of nutrients, water, and oxygen. This accelerated growth can help you produce food more quickly, which is crucial in survival situations.

Space-saving: Hydroponic systems can be designed vertically, allowing you to grow a larger number of plants in a smaller space—perfect for confined shelters or indoor environments.

Basic Components of a Hydroponic System

To get started with hydroponics, you'll need a few essential components. While there are several types of hydroponic systems, most of them rely on the same basic elements to provide plants with the necessary nutrients and growing conditions.

Growing Medium

Although hydroponics doesn't use soil, it does require a growing medium to support the plant roots and retain moisture. Some common growing mediums include:

Rockwool: A fibrous material that provides excellent water retention and root support.

Coco coir: Made from coconut husks, coco coir is a renewable resource that offers good water retention and aeration.

Clay pellets: These lightweight pellets provide good drainage and are easy to reuse.

The growing medium holds the plant in place while allowing the roots to access the nutrient solution.

Nutrient Solution

In a hydroponic system, plants rely on a water-based solution containing all the essential nutrients they need to grow. This nutrient solution typically includes a balanced mix of nitrogen, phosphorus, potassium, calcium, magnesium, and other trace elements. You can purchase pre-made hydroponic nutrient solutions or mix your own using concentrated fertilizers.

Water Reservoir and Pump

The water reservoir holds the nutrient solution and provides it to the plants. Most hydroponic systems use a pump to circulate the water, ensuring that the plants receive a consistent supply of nutrients. The reservoir should be large enough to hold sufficient water for the number of plants you are growing, and it's important to regularly monitor and maintain the water quality.

Grow Lights

If you're growing plants indoors or in a fallout shelter, natural sunlight may not be available or sufficient. Grow lights are essential for providing the light spectrum needed for photosynthesis. LED grow lights are energy-efficient and can be tailored to emit the ideal wavelengths for plant growth, making them a popular choice for hydroponic systems.

pH and EC Meters

Maintaining the correct pH and electrical conductivity (EC) of the nutrient solution is crucial for healthy plant growth. Most plants thrive in a slightly acidic environment with a pH between 5.5 and 6.5. The EC meter measures the concentration of nutrients in the water, helping you ensure that your plants receive the proper balance of minerals.

Types of Hydroponic Systems

There are several different types of hydroponic systems, each with its own advantages and setup requirements. Here are a few popular systems that are well-suited for growing food in a post-nuclear environment:

Deep Water Culture (DWC)

In a Deep Water Culture system, plant roots are suspended directly in a nutrient-rich water solution. An air pump is used to oxygenate the water, ensuring that the roots receive plenty of oxygen while absorbing nutrients. DWC is one of the simplest and most cost-effective hydroponic systems, making it ideal for beginners.

Pros: Simple to set up, low maintenance, and cost-effective.

Cons: Limited plant mobility; not ideal for plants with large root systems.

Nutrient Film Technique (NFT)

The Nutrient Film Technique involves a shallow stream of nutrient solution that flows continuously over the roots of the plants, which are held in a tray or channel. This allows the roots to absorb both water and oxygen efficiently. NFT systems are popular for growing leafy greens and herbs.

Pros: Efficient use of water and nutrients, ideal for smaller plants.

Cons: Requires more precise control of water flow; not suitable for plants with large root systems.

Ebb and Flow (Flood and Drain)

In an Ebb and Flow system, the growing tray is periodically flooded with the nutrient solution, allowing the plants to absorb water and nutrients. The solution is then drained back into the reservoir, allowing the roots to receive oxygen between cycles. This system works well for a variety of vegetables, including tomatoes and peppers.

Pros: Versatile, supports a wide variety of plants.

Cons: Requires careful monitoring to prevent root rot.

Drip System

In a drip system, nutrient solution is delivered directly to the base of each plant through small drip emitters. This system allows for precise control over the amount of water and nutrients each plant receives and is highly efficient for growing larger plants like tomatoes and cucumbers.

Pros: Precise control of nutrient delivery, suitable for large plants.

Cons: More complex setup; emitters can clog if not properly maintained.

Setting Up and Maintaining a Hydroponic System

Once you've selected the type of hydroponic system that best suits your needs, follow these steps to set it up and maintain it for long-term, radiation-free food production.

Select a Location

Choose a location for your hydroponic system that is protected from fallout particles. If you are in a fallout shelter or enclosed space, ensure that the system is positioned near grow lights and has access to clean water and electricity for pumps and lighting.

Assemble the System

Follow the instructions for setting up your chosen hydroponic system. This typically involves assembling the water reservoir, growing trays or containers, pumps, and tubing. Install the grow lights above the plants and position the air pump (if required) in a convenient location.

Prepare the Nutrient Solution

Mix the nutrient solution according to the specific needs of the plants you are growing. Use pH and EC meters to ensure the solution is properly balanced before adding it to the system. Check the pH regularly and adjust as needed to keep it within the optimal range for plant growth.

Plant Your Vegetables

Place your seedlings or seeds into the growing medium, ensuring that the roots are in contact with the nutrient solution. For larger plants, such as tomatoes or peppers, ensure that the system provides enough support for the plants as they grow.

Monitor and Maintain the System

Regularly check the water levels, nutrient concentration, and pH of the system. Replace the nutrient solution every one to two weeks, or as needed, to ensure that the plants receive a steady supply of fresh nutrients. Monitor the plants for signs of growth or stress, adjusting light levels and nutrient concentrations as needed.

What to Grow in Hydroponics

Hydroponics is especially well-suited for growing leafy greens, herbs, and smaller vegetables, but with the right setup, you can grow a wide variety of crops. Here are some popular options for hydroponic gardening:

Lettuce: One of the easiest crops to grow hydroponically, lettuce grows quickly and can be harvested continuously.

Spinach: Another fast-growing leafy green, spinach thrives in hydroponic systems.

Basil: Herbs like basil, cilantro, and mint do well in hydroponics and are great for adding fresh flavors to your meals.

Tomatoes: While tomatoes require more support, they can grow exceptionally well in hydroponic drip systems.

Peppers: Both sweet and hot peppers are well-suited for hydroponic cultivation, as they require minimal space and can thrive with controlled nutrient delivery.

Hydroponics provides a highly efficient and safe method for growing radiation-free food in a post-nuclear environment. By setting up a controlled, soil-free system indoors or in a protected space, you can cultivate a variety of vegetables and herbs without the risk of contamination. With the right equipment, regular maintenance, and proper nutrient management, hydroponics can ensure a steady supply of fresh, healthy food for you and your family in even the most challenging conditions.

Indoor Gardening: The Best Option in Fallout Zones

In the aftermath of a nuclear event, outdoor gardening may no longer be safe due to the risks of radiation and environmental contamination from fallout. In such situations, indoor gardening becomes the best option for growing safe, uncontaminated food. By cultivating plants inside a controlled environment, you can protect them from radioactive particles, heavy metals, and other pollutants, ensuring a steady supply of fresh, healthy food. This chapter will explore the benefits of indoor gardening in fallout zones, the essential components of a successful indoor garden, and practical steps to establish and maintain an indoor garden for long-term survival.

Why Indoor Gardening is the Best Option in Fallout Zones

Indoor gardening offers significant advantages over outdoor cultivation in a fallout zone. By bringing your garden indoors, you can control the environment, protect your crops from radiation and contaminants, and ensure a reliable source of food. Here's why indoor gardening is the optimal solution in a post-nuclear environment:

Protection from radioactive fallout: Outdoor soil and air may be contaminated with fallout particles, which can pose serious health risks if they come into contact with plants. Indoor gardening eliminates this risk by growing food in a clean, enclosed space where it's shielded from radioactive contamination.

Controlled growing conditions: With indoor gardening, you can control critical factors like temperature, humidity, light, and water, optimizing plant growth and minimizing exposure to environmental hazards. This control ensures that plants can thrive even in the harsh conditions of a fallout zone.

Efficient use of space: Indoor gardening allows you to make the most of limited space, especially if you're confined to a shelter or small indoor area. Vertical gardens, shelves, or hydroponic systems enable you to grow more food in a compact environment.

Year-round food production: Indoors, you're not limited by seasonal changes or weather conditions. With the right setup, you can grow food year-round, ensuring a consistent supply of fresh vegetables even in the coldest or most inhospitable climates.

Reduced water consumption: Indoor gardens, particularly those using hydroponic or container-based systems, use significantly less water than traditional outdoor gardening. This is especially important in a survival scenario where access to clean water may be limited.

Setting Up Your Indoor Garden

Creating an indoor garden in a fallout zone requires careful planning and attention to detail. Here are the essential components of a successful indoor gardening system and practical steps to get started:

Choosing the Right Space

The first step in establishing an indoor garden is selecting a suitable space within your shelter or home. The location should be secure, free from contamination, and have enough room for your plants to grow.

Available space: If you're in a fallout shelter or a small indoor space, maximize your available area by using shelves, vertical gardens, or hanging planters. A single wall or corner can be converted into a productive gardening area.

Ventilation: Good air circulation is important for plant health, so ensure your indoor garden has adequate ventilation. Stagnant air can lead to mold growth and pest infestations, so consider using fans to promote airflow.

Access to utilities: You'll need access to electricity for grow lights, water for irrigation, and possibly heat for temperature control. Set up your garden in an area where these resources are easily accessible.

Lighting: The Key to Indoor Gardening

In the absence of natural sunlight, artificial grow lights are essential for indoor gardening. Plants require light to photosynthesize, and the quality of light they receive will determine how well they grow.

LED grow lights: LED lights are the most energy-efficient option for indoor gardening. They emit the specific light wavelengths that plants need for growth, such as blue light for vegetative growth and red light for flowering. LEDs also produce less heat, reducing the risk of overheating your plants.

Fluorescent lights: Fluorescent lights are another affordable option for indoor gardening. They work well for growing smaller plants, herbs, or leafy greens but may not be sufficient for larger, fruiting plants like tomatoes or peppers.

Light cycles: Plants need a balance of light and darkness to grow properly. Typically, vegetables require 12-16 hours of light per day, followed by 8-12 hours of darkness. Use a timer to automate your grow lights and ensure consistent lighting for your plants.

Choosing the Right Containers

Indoor gardening requires the use of containers or grow beds to house your plants. These containers should provide adequate drainage and support for plant roots while conserving space.

Pots and containers: Choose containers that are large enough to accommodate the plant's root system. For larger plants like tomatoes or peppers, use deeper pots (5 gallons or more), while smaller herbs and greens can grow in shallow containers.

Vertical gardening: If space is limited, consider vertical gardening options like hanging planters, wall-mounted shelves, or tiered racks. Vertical gardens maximize your growing area and allow you to cultivate more plants in a smaller space.

Self-watering planters: These planters are ideal for indoor gardening, as they reduce the need for frequent watering and help prevent overwatering. The built-in reservoir system allows plants to absorb water as needed, making maintenance easier.

Soil and Growing Mediums

The type of growing medium you use will depend on whether you're growing plants in traditional containers, using a hydroponic system, or experimenting with other methods like aquaponics. For soil-based indoor gardens, choose clean, uncontaminated soil.

Potting mix: Use a high-quality potting mix that is light, well-draining, and rich in organic matter. Avoid using outdoor soil, as it may be contaminated with fallout particles or pathogens.

Hydroponics: If you prefer a soilless approach, consider using hydroponics (discussed in Chapter 29), where plants grow in a nutrient-rich water solution. Hydroponics is ideal for indoor environments because it eliminates the risk of soil contamination and allows for more efficient use of water and space.

Soil testing: If you're using any soil, test it regularly to ensure it remains uncontaminated and healthy for growing food. Indoor soil can sometimes accumulate salts or develop nutrient imbalances, so it's important to monitor its condition.

Irrigation and Watering Systems

Indoor plants need a consistent supply of clean, uncontaminated water to thrive. Since water in a fallout zone may be scarce, efficient irrigation systems are key to conserving water and ensuring that plants receive the right amount of moisture.

Manual watering: For small-scale indoor gardens, manual watering with a watering can is sufficient. Be careful not to overwater, as this can lead to root rot in container-grown plants.

Drip irrigation: A drip irrigation system delivers water directly to the base of each plant through small emitters, reducing water waste and preventing water from sitting on leaves, which can lead to mold or disease. Drip systems are easy to set up and conserve water.

Self-watering systems: If you are short on time or water, self-watering containers or wicking systems can help keep your plants hydrated with minimal effort. These systems draw water from a reservoir as needed, ensuring that plants always have access to moisture without the risk of overwatering.

Maintaining Proper Temperature and Humidity

Indoor plants thrive in stable environments with consistent temperatures and humidity levels. Depending on the type of plants you're growing, the optimal temperature and humidity will vary, but most vegetables prefer moderate conditions.

Temperature control: Vegetables grow best in temperatures between 65°F and 75°F (18°C to 24°C). If you're in a cold environment or fallout shelter, you may need to use a space heater or heating mats to maintain a warm growing environment. Avoid extreme fluctuations in temperature, as this can stress plants.

Humidity control: Maintaining proper humidity levels (around 50-60%) is important for healthy plant growth. Too much humidity can lead to mold or fungal growth, while too little can cause plants to dry out. Use a humidifier or dehumidifier to adjust the humidity level as needed, and monitor it regularly.

What to Grow Indoors: Best Crops for Indoor Gardening

Certain vegetables and herbs are better suited for indoor growing due to their compact size, fast growth, and ability to thrive in controlled environments. Here are some of the best crops to grow indoors in a fallout zone:

Leafy Greens

Lettuce: Lettuce grows quickly and can be harvested continuously, making it an excellent choice for indoor gardening.

Spinach: Spinach thrives in indoor environments and provides a nutrient-rich addition to your diet.

Kale: Kale is a hardy leafy green that grows well indoors, providing a steady supply of vitamins and minerals.

Herbs

Basil: Basil is easy to grow indoors and adds fresh flavor to your meals.

Mint: Mint is a fast-growing herb that can be grown in pots and adds variety to your indoor garden.

Cilantro: Cilantro grows well in small containers and can be harvested frequently.

Small Vegetables

Tomatoes: Dwarf or cherry tomato varieties are well-suited for indoor growing and can produce abundant fruit in small spaces.

Peppers: Both sweet and hot peppers can thrive indoors, especially in self-watering containers or hydroponic systems.

Cucumbers: Compact cucumber varieties can be grown indoors with the right support, such as trellises or vertical gardening systems.

Microgreens

Broccoli microgreens: These nutrient-dense microgreens grow quickly and are easy to harvest in small indoor spaces.

Radish microgreens: Microgreens like radish can be grown in shallow trays and harvested within a few weeks, providing a fast source of fresh food.

Maintaining and Expanding Your Indoor Garden

Once your indoor garden is up and running, maintaining it will involve regular monitoring and adjustments to ensure that plants remain healthy. Here are some tips for keeping your indoor garden productive over the long term:

Monitor Light and Water Levels

Regularly check that your grow lights are positioned correctly and providing enough light for the plants. Ensure your watering system is functioning properly, and avoid overwatering or letting plants dry out.

Prune and Harvest Regularly

To encourage continuous growth, prune your plants and harvest regularly. This helps prevent overcrowding, improves airflow, and stimulates the production of new leaves or fruit.

Replenish Nutrients

If you're growing in containers or using hydroponics, replenish the nutrients in the soil or water solution periodically. Indoor plants can deplete nutrients faster, so monitor their growth and adjust the nutrient mix as needed.

Expand Your Garden

As you gain experience, consider expanding your indoor garden by adding more plants, experimenting with new crops, or introducing hydroponic systems. Vertical gardening can help you maximize your growing space and increase your food supply.

Indoor gardening is the best option for growing safe, uncontaminated food in a fallout zone. By controlling the environment and using efficient systems for lighting, irrigation, and temperature management, you can cultivate a wide variety of crops in a compact space. With the right setup, you can grow fresh vegetables and herbs year-round, ensuring a sustainable and reliable source of food for long-term survival. Indoor gardening not only helps you avoid the risks of radioactive contamination but also provides a practical solution for producing food in even the harshest conditions.

Nuclear Winter: What It Is and What to Expect

A nuclear winter is one of the most catastrophic and far-reaching consequences of a large-scale nuclear conflict. While the immediate effects of nuclear explosions—such as blasts, fires, and radiation—are devastating, the aftermath of widespread nuclear detonations can lead to long-term environmental destruction on a global scale. A nuclear winter occurs when massive amounts of smoke, soot, and dust from burning cities, forests, and other materials rise into the upper atmosphere, blocking sunlight and drastically lowering temperatures across the world. This chapter will explain what nuclear winter is, the science behind it, its potential impacts on the environment and human survival, and what you can expect if such an event occurs.

What is Nuclear Winter?

Nuclear winter refers to the severe and prolonged global climatic cooling effect that follows a large-scale nuclear war. The phenomenon occurs because multiple nuclear explosions ignite widespread fires, particularly in cities and industrial areas, releasing vast quantities of smoke, ash, and soot into the atmosphere. This particulate matter, particularly the black carbon from burning materials, absorbs sunlight and prevents it from reaching the Earth's surface, leading to a dramatic drop in temperatures.

The term "nuclear winter" was coined in the 1980s by scientists who developed models to predict the potential environmental effects of nuclear war. These models suggested that the combination of darkness, cold, and radioactive fallout could lead to catastrophic disruptions in the Earth's ecosystems, agriculture, and human societies.

The Science behind Nuclear Winter

To understand nuclear winter, it's important to grasp the basic mechanisms that would lead to such a global environmental change:

Soot and Smoke Injection into the Atmosphere: When nuclear weapons detonate over cities, forests, and industrial areas, the resulting fires (firestorms) produce immense amounts of black carbon soot. This soot is carried by rising hot air into the stratosphere, the second layer of Earth's atmosphere.

Blocking of Sunlight: Once in the stratosphere, the soot and smoke particles form a thick layer that absorbs and scatters sunlight, preventing much of it from reaching the Earth's surface. This causes a significant reduction in solar radiation, leading to a drop in temperatures.

Cooling of the Earth's Surface: The blocked sunlight causes surface temperatures to fall rapidly. Depending on the scale of the nuclear war, temperatures could drop by several degrees Celsius (or Fahrenheit), creating winter-like conditions even during summer months. The cooling would likely be most severe in temperate and high-latitude regions, but the effects would be felt worldwide.

Disruption of Weather Patterns: The cooling of the Earth's surface would disrupt atmospheric circulation and weather patterns. This could lead to longer winters, shorter growing seasons, and altered precipitation patterns. Some areas might experience prolonged droughts, while others could see increased rainfall.

Persistence of Effects: The soot and smoke particles in the stratosphere could remain there for years, as there is little weather (rain or wind) in this layer of the atmosphere to wash them out. This means the effects of nuclear winter could last for a decade or more, even if the war itself only lasted for a short time.

Potential Environmental Impacts of Nuclear Winter

The environmental consequences of nuclear winter would be severe and far-reaching, affecting every aspect of life on Earth. Here are the key impacts you can expect:

Global Cooling

The most immediate effect of nuclear winter is the sharp drop in global temperatures. In the worst-case scenario, temperatures could fall by 10 to 20 degrees Celsius (18 to 36 degrees Fahrenheit), creating an artificial winter that lasts for years. Even a smaller nuclear conflict could result in a temperature drop of 2 to 5 degrees Celsius (4 to 9 degrees Fahrenheit), which would still have devastating effects on the climate.

Length of cooling: The cooling effect could last for years or even decades, depending on how much soot is injected into the atmosphere and how long it takes for natural processes to remove it.

Global consequences: The cooling would not be evenly distributed; temperate regions, such as North America, Europe, and parts of Asia, would experience the most dramatic drops in temperature. However, tropical and subtropical regions would also cool, leading to widespread changes in weather and ecosystems.

Severe Agricultural Disruptions

Agriculture would be one of the hardest-hit sectors during a nuclear winter. The combination of lower temperatures, reduced sunlight, and changes in precipitation would make it extremely difficult to grow crops, leading to global food shortages.

Shortened growing seasons: In many parts of the world, the growing season would become too short to produce staple crops like wheat, corn, rice, and soybeans. Frosts could occur during traditionally warm months, killing plants before they mature.

Decline in crop yields: Even in areas where crops could still be grown, reduced sunlight and colder temperatures would drastically lower yields. Plants need sunlight to photosynthesize, and without it, they cannot grow properly. Yields for most crops would likely fall by 50% or more, leading to widespread famine.

Livestock losses: Livestock would also suffer during a nuclear winter, as they depend on crops and grasses for food. With reduced agricultural output, farmers would be unable to feed their animals, leading to mass die-offs of cattle, sheep, pigs, and other domesticated animals.

Widespread Famine and Food Insecurity

With global agricultural systems collapsing, famine would become a severe problem. Countries that rely on imports for food, as well as regions that are already food-insecure, would be hit the hardest. Even nations with traditionally strong agricultural sectors would struggle to feed their populations due to the drastic reductions in crop yields.

Global food shortages: The combination of failed crops, lost livestock, and reduced global food trade would lead to worldwide food shortages. The price of food would skyrocket, and governments would likely impose rationing to control the distribution of available supplies.

Mass starvation: In the worst-case scenario, mass starvation could occur, particularly in densely populated urban areas and developing nations. Some estimates suggest that billions of people could die from starvation in the years following a nuclear winter.

Disruption of Ecosystems

The dramatic cooling and loss of sunlight would cause widespread disruptions to ecosystems around the world. Many species would struggle to survive in the altered climate, leading to mass extinctions.

Loss of plant life: The reduction in sunlight would cause forests, grasslands, and other plant communities to wither. This would have a cascading effect on the animals that depend on those plants for food, leading to a collapse of ecosystems.

Marine life disruption: The cooling of surface waters would also affect marine ecosystems. The ocean's ability to absorb carbon dioxide could be reduced, altering marine chemistry and threatening species like plankton, which form the base of the marine food chain. This would have ripple effects throughout the oceanic ecosystem, impacting fish, marine mammals, and seabirds.

Extinction of species: Species that are sensitive to changes in temperature or food availability would be at risk of extinction. This could include many types of birds, insects, and mammals, particularly those in temperate regions.

Human Survival in a Nuclear Winter

Surviving in a nuclear winter would be incredibly difficult. The combination of cold temperatures, food shortages, and the breakdown of infrastructure would create a hostile environment where resources are scarce and life is precarious. Here's what to expect if you're trying to survive a nuclear winter:

Severe Cold and Shelter Needs

The sharp drop in temperatures would make it difficult to stay warm, especially for those without access to adequate heating and shelter. Many homes and buildings would be ill-prepared to handle freezing temperatures over long periods, and the energy grid may be damaged or overwhelmed.

Heating sources: Staying warm would require reliable heating sources. Wood stoves, fireplaces, and generators could provide heat, but fuel sources might become scarce as infrastructure breaks down. If the power grid collapses, electrical heating systems would be useless.

Layering and insulation: Wearing layers of clothing, using blankets, and insulating your shelter as much as possible would be necessary to retain heat. Underground shelters or well-insulated homes would provide better protection from the cold.

Food and Water Scarcity

With agriculture disrupted and food supplies dwindling, ensuring a reliable food source would be one of the biggest challenges in a nuclear winter. Preppers and survivalists who have stockpiled food would be better equipped to survive, but even these supplies may run out over time.

Preserved food: Stockpiling non-perishable food like canned goods, dried foods, and freeze-dried meals would be essential for surviving the initial months or years of nuclear winter. Learning how to preserve food through canning, drying, and fermentation could also help prolong your food supply.

Water filtration: Clean water would also be a critical concern, as fallout could contaminate natural water sources. Water filtration and purification systems would be essential for ensuring safe drinking water.

Social Collapse and Security Concerns

In a nuclear winter scenario, societal structures may collapse due to resource scarcity, political instability, and mass migration. Desperate people could turn to looting, violence, and other forms of crime as they struggle to survive, creating dangerous conditions for those trying to maintain their own safety.

Community survival: Forming strong, self-sustaining communities could be key to surviving in the long term. Sharing resources, knowledge, and skills with others could improve your chances of survival while reducing the risk of isolation or vulnerability.

Security and defense: Defending yourself and your community from theft, looting, and violence would be a significant concern. Preparing for potential conflicts and having a strategy for securing your resources would be critical in maintaining safety during a societal breakdown.

A nuclear winter would be one of the most devastating global consequences of nuclear war, with long-lasting effects on the environment, food production, and human survival. The sudden and severe drop in temperatures, combined with the collapse of agriculture and ecosystems, would create a world of scarcity and hardship. While surviving a nuclear winter is possible with the right preparation—such as securing shelter, food, and water—it would be an incredibly challenging ordeal that requires careful planning, resilience, and adaptability. Understanding what to expect and how to prepare for such a catastrophic event is essential for anyone considering long-term survival in a post-nuclear world.

Understanding Radiation Decay: When Will It Be Safe?

After a nuclear event, one of the most critical questions for survival is understanding when the environment will be safe from radiation. Radioactive fallout can contaminate the air, soil, water, and living organisms, making it dangerous to go outside, grow food, or drink water without proper precautions. However, radiation doesn't remain at dangerous levels forever—it decays over time. Understanding the process of radiation decay and knowing when it will be safe to return to normal activities is crucial for long-term survival. In this chapter, we'll explore how radiation decays, the types of radioactive isotopes you may encounter, and how to estimate when it will be safe to re-enter contaminated areas.

What is Radiation Decay?

Radiation decay, also known as radioactive decay, is the process by which unstable atomic nuclei lose energy by emitting radiation. This process occurs naturally as radioactive elements, known as radionuclides, break down over time. The rate of decay is measured by a substance's half-life—the time it takes for half of the radioactive atoms in a material to decay into a stable form.

Each radioactive isotope has a different half-life, ranging from fractions of a second to millions of years. Understanding the half-lives of the radioactive isotopes present in fallout is essential for determining how long an area will remain dangerous and when it will be safe to return.

Types of Radioactive Isotopes in Fallout

Nuclear explosions produce a variety of radioactive isotopes, some of which have short half-lives, while others persist in the environment for much longer. The following are some of the most common radioactive isotopes found in nuclear fallout:

Iodine-131

Half-life: 8 days

Decay time: 99% decayed in 80 days

Health risks: Iodine-131 is rapidly absorbed by the thyroid gland and can lead to thyroid cancer. It is one of the most dangerous isotopes immediately after a nuclear event.

Protection: Taking potassium iodide (KI) can block the thyroid from absorbing radioactive iodine, reducing the risk of thyroid cancer.

Cesium-137

Half-life: 30 years

Decay time: 99% decayed in 600 years

Health risks: Cesium-137 is a long-lived isotope that contaminates soil and water, where it can be absorbed by plants and animals. It emits beta and gamma radiation, which can lead to radiation sickness, cancer, and genetic damage.

Protection: Avoid consuming contaminated food and water. Cesium can spread through ecosystems, so it is essential to test and avoid affected areas for decades after a nuclear event.

Strontium-90

Half-life: 28.8 years

Decay time: 99% decayed in 576 years

Health risks: Strontium-90 behaves like calcium and accumulates in bones and teeth, where it emits beta radiation. It increases the risk of bone cancer and leukemia.

Protection: Prevent consumption of contaminated food and water, particularly milk and dairy products, as strontium-90 can be absorbed by animals and passed to humans through these products.

Plutonium-239

Half-life: 24,100 years

Decay time: 99% decayed in 482,000 years

Health risks: Plutonium-239 is highly toxic if inhaled or ingested. It emits alpha particles, which can cause lung cancer, liver damage, and bone cancer if it enters the body. Fortunately, alpha particles cannot penetrate the skin, so external exposure is less of a concern.

Protection: Avoid inhaling or ingesting plutonium particles, and stay clear of areas contaminated with plutonium for the foreseeable future.

How Radiation Decays Over Time

Radiation levels do not decrease evenly but follow a pattern based on the half-lives of the radioactive isotopes present. This means that radiation levels fall rapidly at first and then decline more slowly over time. The general rule is known as the "7-10 Rule": for every sevenfold increase in time after a nuclear explosion, the radiation levels decrease by a factor of 10.

First hour: After the initial blast, radiation levels are extremely high.

After 7 hours: Radiation levels will have dropped to 10% of their original levels.

After 49 hours (2 days): Radiation levels will be 1% of their original levels.

After 14 days: Radiation levels will be about 0.1% of the initial levels.

After 3 months: Radiation levels may drop to 0.01% of their initial levels, depending on the type of fallout.

However, while this decay reduces the immediate danger, longer-lived isotopes like cesium-137 and strontium-90 will continue to pose risks for decades to centuries, contaminating soil, water, and food sources.

Estimating When It Will Be Safe

Determining when it will be safe to re-enter fallout-affected areas depends on several factors, including the type of radioactive isotopes present, the intensity of the initial radiation levels, and the duration of exposure. Here are some general guidelines for assessing when it will be safe to return to a fallout zone:

Short-Term Safety (First Few Days)

In the first few days following a nuclear explosion, the radiation levels are extremely high due to short-lived isotopes like iodine-131. Staying indoors or in a fallout shelter is essential during this time. It is generally not safe to go outside for more than a few minutes in the first 48 hours.

Safe activities: Remain sheltered for at least 24 to 48 hours after the blast. Only leave the shelter for short periods, and only if necessary, wearing protective gear to minimize exposure.

Medium-Term Safety (First Few Weeks)

After the first two weeks, radiation levels from short-lived isotopes like iodine-131 will have decayed significantly. However, isotopes like cesium-137 and strontium-90 will still pose a serious threat. If radiation levels have dropped to safer levels (typically below 0.5 microsieverts per hour), limited outdoor activities may be possible.

Safe activities: At this point, it may be possible to leave the shelter for longer periods, but caution is still required. Avoid consuming any food or water from contaminated areas, and test the environment regularly with a Geiger counter.

Long-Term Safety (Months to Years)

In the months and years following a nuclear event, radiation levels will continue to decay, but long-lived isotopes like cesium-137 and strontium-90 will remain hazardous for decades. Returning to normal agricultural activities or consuming food from fallout-affected areas may not be safe for many years, depending on local contamination levels.

Safe activities: Long-term survival requires careful monitoring of radiation levels. Avoid areas with high contamination, particularly those with persistent isotopes. Indoor activities, hydroponic gardening, or using uncontaminated soil for growing food are recommended until the environment becomes safer.

Tools for Measuring Radiation Decay

To track radiation decay and assess when it will be safe to re-enter contaminated areas, you'll need reliable tools for measuring radiation. These devices allow you to determine the intensity of radiation in your environment and make informed decisions about safety.

Geiger Counter

A Geiger counter is the most commonly used device for measuring radiation. It detects ionizing radiation (such as alpha, beta, and gamma radiation) and provides real-time readings of radiation levels in the environment. Geiger counters are essential for monitoring radiation decay and assessing whether an area is safe to enter.

Dosimeter

A dosimeter measures the cumulative exposure to radiation over time. This tool is particularly useful for tracking how much radiation you or your family members have been exposed to while outdoors or working in contaminated areas. Personal dosimeters can be worn on the body and checked periodically.

Radiation Detection Badges

Radiation detection badges work similarly to dosimeters but are usually worn for longer periods (e.g., days or weeks) to track exposure. These badges change color based on the amount of radiation absorbed, offering a simple visual indicator of whether you've been exposed to dangerous levels of radiation.

Factors that Influence Radiation Decay

While the half-lives of radioactive isotopes give a good estimate of when radiation levels will decrease, other factors can influence how quickly an area becomes safe. These include:

Weather: Rain and wind can wash radioactive particles into the ground or water, affecting how long they remain in the atmosphere. However, this can also lead to concentrated pockets of radiation in certain areas (e.g., "hot spots").

Type of terrain: Soil and landscape features, such as mountains, valleys, and water bodies, can influence how fallout settles. Low-lying areas may trap fallout, while wind may disperse radioactive particles more quickly in open plains.

Human intervention: In some cases, decontamination efforts, such as removing topsoil or cleaning up debris, can help accelerate the recovery of an area by reducing the concentration of radioactive particles.

Radiation decay is a slow but steady process, and understanding how it works is key to knowing when it will be safe to re-enter fallout zones. While short-lived isotopes like iodine-131 decay relatively quickly, long-lived isotopes like cesium-137 and strontium-90 can persist in the environment for decades, posing long-term risks. Monitoring radiation levels with tools like Geiger counters and dosimeters, and following general guidelines for short-, medium-, and long-term safety, will help you make informed decisions about when it's safe to resume outdoor activities, grow food, and return to affected areas. Ultimately, surviving in a post-nuclear world requires patience, vigilance, and a clear understanding of how radiation decays over time.

Nuclear Aftermath: How Society Might Rebuild

After a nuclear event, society would face immense challenges in rebuilding and restoring some semblance of normalcy. A large-scale nuclear disaster would likely destroy infrastructure, disrupt communication, collapse economies, and result in widespread loss of life, all of which could lead to political instability and social upheaval. Despite the enormity of these challenges, history has shown that human societies are resilient, capable of rebuilding even after catastrophic events. In this chapter, we will explore how society might rebuild in the aftermath of a nuclear disaster, focusing on infrastructure, governance, community organization, economic recovery, and the psychological and cultural shifts that would likely emerge.

The Initial Impact: A Broken World

The immediate aftermath of a nuclear event would be marked by devastation on an unprecedented scale. Depending on the size and scope of the conflict, millions could die in the initial explosions and from radiation exposure, with many more suffering from long-term health issues such as radiation sickness and cancer. Entire cities could be leveled, and critical infrastructure—power grids, transportation systems, hospitals, and food distribution networks—would be destroyed or severely damaged. In the wake of such destruction, society would enter a period of survival and adaptation before meaningful rebuilding could begin.

The steps toward rebuilding would depend on several factors, including the scale of the nuclear event, the extent of the damage, and the availability of resources and leadership. In this environment, communities would need to find ways to sustain themselves, re-establish order, and work together to create a foundation for future growth.

Rebuilding Infrastructure

One of the first tasks in the aftermath of a nuclear disaster would be to begin rebuilding the basic infrastructure needed for survival. This includes shelter, transportation, energy, water, and sanitation systems. Without these essential elements, any attempt to rebuild society would be unsustainable.

Shelter and Housing

The destruction of homes and buildings would leave millions displaced. Initially, survivors would need to rely on temporary shelters, such as fallout shelters, tents, or repurposed buildings that survived the blast. Over time, more permanent housing solutions would need to be constructed.

Rebuilding homes: Communities would begin by using available resources—such as salvaged materials from the ruins of cities or natural materials like wood and stone—to rebuild homes. Modular and prefabricated housing could become important, as these structures can be built quickly and with fewer materials.

Underground shelters: In areas with continued radiation risks, underground shelters or bunkers might serve as long-term housing solutions. These structures offer protection from lingering radiation and extreme weather, both of which could be concerns in the wake of a nuclear disaster.

Transportation Networks

Roads, bridges, and railways would likely be destroyed or impassable due to fallout or debris. Re-establishing transportation networks would be critical for moving people, resources, and supplies during the rebuilding process.

Clearing debris: The first step would involve clearing debris and making roads passable again. This might require basic tools and manpower before heavy machinery can be reintroduced.

Repairing essential routes: Rebuilding key transportation routes to connect surviving communities would be a priority. These routes would allow for the distribution of food, water, and medical supplies, as well as facilitate trade and communication between regions.

Energy and Power

The electrical grid would likely be one of the most damaged parts of the infrastructure, with power plants destroyed or rendered inoperable. Rebuilding energy production and distribution systems would take time, and in the interim, communities would need to rely on alternative power sources.

Alternative energy: Solar panels, wind turbines, and small-scale hydroelectric generators could be used to generate power in the short term. Solar panels, in particular, are portable and scalable, making them suitable for communities trying to re-establish power.

Localized grids: Rebuilding large, centralized power grids might not be feasible in the immediate aftermath of a nuclear event. Instead, communities could focus on building localized microgrids that are more resilient and easier to repair. These grids could connect homes, hospitals, and essential services to renewable energy sources.

Water and Sanitation

Clean water would be one of the most urgent needs after a nuclear disaster, as many natural water sources could be contaminated by fallout. Rebuilding water infrastructure would involve both securing clean water sources and re-establishing sanitation systems to prevent the spread of disease.

Water filtration and purification: Initially, survivors would need to rely on water filtration systems, such as portable filters or distillation devices, to make contaminated water safe to drink. Communities would also need to identify uncontaminated groundwater sources, such as wells or springs.

Rebuilding water infrastructure: Over time, communities would need to repair damaged water treatment facilities and pipes. In the absence of these systems, communities might develop rainwater collection systems or small-scale purification plants to ensure a reliable water supply.

Communication Systems

Restoring communication would be crucial for coordinating relief efforts, sharing information, and re-establishing governance. However, the destruction of telecommunications infrastructure would make this challenging.

Shortwave radios: In the immediate aftermath, survivors would likely rely on shortwave radios for communication. These radios can operate over long distances and don't require large infrastructure to function.

Rebuilding telecommunications: Over time, communities would work to rebuild telecommunications networks, starting with simple technologies like landlines and progressing to more advanced systems like cell towers and internet services.

Re-establishing Governance and Law

A nuclear disaster could leave many governments in disarray, particularly if major political and economic centers are destroyed. In the absence of central authority, local communities would need to create their own systems of governance to maintain order and coordinate rebuilding efforts.

Community Leadership

In the absence of national governments, local leaders would likely emerge to fill the power vacuum. These leaders could be individuals with survival skills, military training, or experience in crisis management. Community leadership would be essential in organizing recovery efforts, distributing resources, and maintaining security.

Localized governments: Communities might establish councils or assemblies to govern themselves. These local governments would be responsible for making decisions about resource allocation, security, and rebuilding.

Decentralized governance: Given the challenges of communication and travel in a post-nuclear world, governance would likely become decentralized, with local governments having more autonomy over their affairs. This could lead to the formation of regional or city-states, each with its own laws and leadership.

Law and Order

Maintaining law and order would be a significant challenge in a post-nuclear society. With widespread destruction and resource scarcity, crime and violence could increase as people struggle to survive. Communities would need to establish security forces or militias to protect against looting, theft, and violence.

Community policing: In the absence of formal law enforcement, communities may rely on volunteer militias or neighborhood watch groups to maintain security. These groups would enforce local laws and protect resources from looters or criminal elements.

Justice systems: Rebuilding formal justice systems, such as courts and legal codes, would take time. Initially, communities might resolve disputes through mediation or community councils. Over time, more formal systems of justice could be reintroduced as society stabilizes.

Reviving the Economy

A nuclear disaster would devastate the global economy, disrupting trade, destroying industries, and devaluing currencies. Rebuilding the economy would be a long and difficult process, but it would be essential for restoring a functional society.

Bartering and Resource-Based Economies

In the early stages of recovery, the formal economy might collapse, with money becoming useless due to hyperinflation or loss of trust in currency. In this context, bartering and resource-based economies would emerge, with goods and services exchanged directly.

Barter systems: Communities would likely establish barter systems, trading goods like food, water, fuel, and tools. Skills and services, such as carpentry, medical care, or mechanics, could also become valuable forms of currency.

Precious metals and goods: Over time, precious metals like gold and silver, as well as essential goods like fuel or medical supplies, might become accepted as currency. These items hold intrinsic value and could serve as a medium of exchange in the absence of formal currencies.

Rebuilding Trade Networks

Trade would be critical to recovery, as communities would need access to resources that are not available locally. Re-establishing trade networks, both local and regional, would help distribute goods and rebuild the economy.

Local trade: Initially, trade would be limited to nearby communities, as transportation networks would still be under repair. Goods like food, water, and raw materials would be exchanged between towns and villages.

Regional trade: As transportation networks are rebuilt and security improves, trade between regions would resume. Resources like fuel, metals, and machinery would be essential for rebuilding infrastructure and industry.

Rebuilding Industry

The destruction of factories, power plants, and manufacturing centers would cripple industrial production in the short term. However, over time, communities would begin rebuilding small-scale industries to produce essential goods.

Small-scale manufacturing: Initially, communities might focus on small-scale manufacturing, using simple tools and local materials to produce basic goods like clothing, tools, and building materials. Over time, larger industries could be rebuilt, especially if there is access to renewable energy sources.

Agriculture: Reviving agriculture would be a priority, as food production is essential for long-term survival. In the absence of large-scale farms, communities might focus on local, sustainable farming methods, such as indoor gardening, hydroponics, or raising livestock.

Cultural and Psychological Shifts

A nuclear disaster would profoundly affect the way people think, act, and view the world. The trauma of surviving such an event, combined with the loss of familiar social structures, would lead to significant psychological and cultural changes.

Psychological Resilience

Surviving the initial shock of a nuclear event would require immense psychological resilience. Many survivors would struggle with grief, fear, and uncertainty as they try to rebuild their lives in a changed world.

Mental health: Addressing mental health issues would be essential for helping people cope with the trauma of the disaster. Communities might establish support groups, encourage social interaction, and provide counseling services to help survivors deal with anxiety, depression, and PTSD.

Cultural shifts: The experience of living through a nuclear disaster could lead to a cultural shift toward greater self-reliance, cooperation, and preparedness. Survivors might place more value on skills like farming, carpentry, and medicine, as well as on community ties and mutual support.

Education and Knowledge Preservation

Rebuilding society would require not only physical resources but also knowledge. Preserving and passing on essential knowledge about agriculture, medicine, engineering, and other fields would be critical for long-term recovery.

Survival skills education: In the short term, education might focus on teaching survival skills, such as growing food, treating injuries, and building shelters. Communities could establish schools or workshops to pass on this knowledge to younger generations.

Preserving knowledge: Libraries, universities, and archives might be destroyed in a nuclear event, but preserving knowledge for future generations would be vital. Communities could prioritize saving books, digital files, and other records that contain essential information for rebuilding society.

Rebuilding society after a nuclear disaster would be an enormous challenge, requiring resilience, cooperation, and innovation. The process would likely begin at the local level, with communities working together to restore basic infrastructure, governance, and economic systems.

Radiation and the Human Body: Long-Term Effects on Health

Radiation exposure is one of the most serious and potentially life-threatening consequences of a nuclear event. While the immediate effects of high doses of radiation, such as acute radiation sickness, are well understood, the long-term health impacts of radiation exposure are complex and can manifest months, years, or even decades after the initial exposure. In this chapter, we will explore how radiation affects the human body over the long term, the types of radiation that pose the greatest risks, the common health problems associated with radiation exposure, and strategies for mitigating these effects in a post-nuclear environment.

Types of Radiation and How They Affect the Body

Radiation is energy that travels in the form of waves or particles. In the context of a nuclear event, the most dangerous types of radiation are ionizing radiation, which has enough energy to remove tightly bound electrons from atoms, creating ions. Ionizing radiation can damage or destroy cells, leading to a range of health problems, from burns and tissue damage to cancer and genetic mutations.

There are several types of ionizing radiation that survivors of a nuclear event may encounter:

Alpha Radiation

What it is: Alpha particles are large, positively charged particles that cannot penetrate human skin. However, if inhaled or ingested, alpha particles can cause severe damage to internal organs and tissues.

Health risks: While external exposure to alpha radiation is not harmful, internal exposure can lead to lung, liver, and bone damage. Plutonium-239 is a notable alpha emitter and is highly toxic if inhaled or ingested.

Beta Radiation

What it is: Beta particles are smaller than alpha particles and can penetrate the skin but are generally stopped by clothing or a few millimeters of materials like plastic or glass.

Health risks: Beta radiation can cause skin burns if the skin is exposed to high levels for an extended period. If ingested or inhaled, beta particles can damage internal tissues and increase the risk of cancer. Strontium-90, a beta emitter, can be absorbed by bones, where it continues to emit radiation, increasing the risk of bone cancer.

Gamma Radiation

What it is: Gamma rays are highly penetrating electromagnetic waves that can pass through the human body, causing widespread damage to tissues and organs.

Health risks: Gamma radiation is extremely dangerous and can cause both immediate and long-term health effects. It is associated with radiation sickness, cancer, and DNA damage, leading to genetic mutations. Cesium-137 and iodine-131 are common gamma emitters in nuclear fallout.

Neutron Radiation

What it is: Neutron radiation consists of neutrons that are emitted during nuclear fission. These particles are highly penetrating and can travel through materials, including the human body.

Health risks: Neutron radiation can cause severe tissue damage and increases the risk of cancer. It is primarily a concern for those close to a nuclear explosion or nuclear reactors.

How Radiation Affects the Body

The effects of radiation on the human body depend on several factors, including the dose of radiation received, the duration of exposure, and whether the exposure was internal (through inhalation or ingestion) or external. The body's cells are especially vulnerable to radiation because it damages DNA, leading to mutations, cell death, or uncontrolled cell growth (cancer). Here are some of the key ways radiation affects the body:

Cellular Damage

Radiation damages the DNA inside cells, which can result in mutations or cell death. If the damage is severe, the cell may die or become unable to function properly. If the damage is moderate, the cell may survive but with mutated DNA, which can lead to uncontrolled cell growth and cancer.

Tissue and Organ Damage

The effects of radiation exposure are most noticeable in rapidly dividing tissues, such as the skin, bone marrow, and gastrointestinal tract. These tissues are more sensitive to radiation because they are constantly generating new cells. High doses of radiation can destroy these tissues, leading to acute radiation sickness (ARS), which includes symptoms like nausea, vomiting, diarrhea, and bleeding.

Bone marrow: Radiation damages bone marrow, which produces blood cells. This can lead to a condition known as bone marrow suppression, resulting in reduced production of red blood cells (anemia), white blood cells (immunosuppression), and platelets (increased bleeding).

Gastrointestinal tract: The cells lining the intestines and stomach divide rapidly, making them vulnerable to radiation damage. This can lead to nausea, vomiting, diarrhea, and increased risk of infection.

Cancer

Cancer is one of the most well-documented long-term effects of radiation exposure. The risk of developing cancer increases with the dose of radiation received and the length of time since exposure. Radiation exposure can cause various types of cancer, including:

Leukemia: Leukemia is one of the most common cancers associated with radiation exposure. It occurs when radiation damages the bone marrow, leading to the uncontrolled growth of white blood cells.

Thyroid cancer: Radioactive iodine (iodine-131) released during nuclear explosions is absorbed by the thyroid gland, increasing the risk of thyroid cancer. Potassium iodide (KI) pills can reduce this risk by saturating the thyroid with stable iodine.

Lung cancer: Inhalation of radioactive particles, particularly alpha emitters like plutonium-239, increases the risk of lung cancer.

Breast cancer: Women exposed to radiation, especially at a young age, have an increased risk of developing breast cancer.

Bone cancer: Strontium-90, which behaves like calcium, can accumulate in bones and increase the risk of bone cancer.

Genetic Damage and Birth Defects

Radiation can cause mutations in reproductive cells, leading to genetic damage that can be passed on to future generations. Children born to parents who were exposed to high doses of radiation may be at increased risk of birth defects, developmental issues, and genetic disorders.

In-utero exposure: Pregnant women exposed to radiation are at particular risk, as developing fetuses are highly sensitive to radiation. In-utero exposure can lead to developmental delays, low birth weight, or physical deformities. The severity of the effects depends on the stage of pregnancy and the dose of radiation.

Long-Term Health Effects of Radiation Exposure

The long-term health effects of radiation exposure can take years or even decades to manifest. While some people may recover from radiation exposure without significant health issues, others may develop serious conditions later in life. Here are some of the most common long-term effects of radiation exposure:

Radiation-Induced Cancers

As mentioned, radiation exposure significantly increases the risk of developing cancer. The latency period for radiation-induced cancers can be long, often taking 10 to 20 years or more to develop. Common cancers caused by radiation include leukemia, thyroid cancer, lung cancer, breast cancer, and bone cancer. The risk of cancer increases with higher radiation doses and prolonged exposure.

Cardiovascular Disease

Radiation exposure, especially high doses, can damage the heart and blood vessels, leading to an increased risk of cardiovascular disease later in life. Studies of survivors of the Hiroshima and Nagasaki bombings, as well as patients who received radiation therapy for cancer, have shown an increased risk of heart disease and stroke many years after exposure.

Cataracts

Radiation exposure can damage the lens of the eye, leading to cataracts, a condition in which the lens becomes cloudy and impairs vision. Cataracts are a well-documented long-term effect of radiation exposure, and the risk increases with higher doses. Cataracts can develop several years after exposure and may require surgical intervention.

Chronic Fatigue and Immune System Suppression

Radiation exposure can lead to long-term damage to the immune system, making survivors more vulnerable to infections and diseases. Chronic fatigue is another common symptom, as the body struggles to repair radiation-induced damage to tissues and organs. This fatigue can persist for years after exposure and may be accompanied by weakness, anemia, and other signs of immune system dysfunction.

Mitigating Long-Term Effects of Radiation Exposure

While it's impossible to eliminate all risks associated with radiation exposure, there are steps you can take to mitigate the long-term health effects and improve your chances of survival in a post-nuclear environment:

Minimize Exposure

The best way to reduce the risk of long-term health problems is to minimize your exposure to radiation. This can be done by staying in a fallout shelter for as long as possible after a nuclear event, avoiding contaminated food and water, and using personal protective equipment (PPE) like masks and clothing when going outside.

Take Potassium Iodide (KI)

Potassium iodide (KI) is a key preventive measure that can help protect your thyroid from absorbing radioactive iodine, reducing the risk of thyroid cancer. KI should be taken shortly before or after exposure to radioactive iodine to be effective. Keep in mind that KI does not protect against other types of radiation.

Consume a Nutrient-Rich Diet

A healthy diet rich in antioxidants, vitamins, and minerals can help the body repair radiation-induced damage and support immune function. Foods high in antioxidants, such as fruits and vegetables, may help reduce oxidative stress caused by radiation exposure. Ensuring adequate intake of calcium and potassium can also help reduce the absorption of radioactive isotopes like strontium-90 and cesium-137.

Regular Health Monitoring

If you've been exposed to radiation, it's important to undergo regular health check-ups to monitor for signs of radiation-induced illnesses, particularly cancer. Early detection and treatment of cancers and other radiation-related conditions can significantly improve survival rates.

Cancer screenings: Regular screenings for cancer, especially for high-risk areas like the thyroid, lungs, and bones, are essential for early detection.

Blood tests: Blood tests can monitor the health of your bone marrow, immune system, and overall blood cell count, helping detect issues like anemia or immune suppression early on.

Psychological Support

Survivors of nuclear events often experience long-term psychological effects, including anxiety, depression, and post-traumatic stress disorder (PTSD). Coping with the trauma of radiation exposure and the uncertainty of long-term health risks can take a toll on mental health. Seeking psychological support and building strong community ties can help mitigate the mental health impact of a nuclear event.

Radiation exposure poses significant long-term health risks, ranging from cancer and cardiovascular disease to genetic damage and immune system suppression. Understanding how radiation affects the body and taking steps to mitigate its effects can improve your chances of survival and long-term health in a post-nuclear environment. By minimizing exposure, taking preventive measures like potassium iodide, and maintaining regular health monitoring, you can reduce the risks associated with radiation and better prepare for life after a nuclear event.

Medical Preparations: Stocking Up for Survival

When preparing for survival in a post-nuclear world, medical preparedness is one of the most critical aspects of long-term survival. In a fallout scenario, access to medical professionals, hospitals, and pharmacies may be limited or completely unavailable. Having a well-stocked medical kit and knowledge of basic medical care can make the difference between life and death. In this chapter, we'll cover the essential medical supplies to stock up on, how to prepare for common health issues in a fallout zone, and the basic skills needed to provide medical care for yourself and others.

Why Medical Preparation is Critical

In the aftermath of a nuclear disaster, injuries, illnesses, and radiation-related health problems will be widespread. Medical facilities may be destroyed or overwhelmed, and the supply chain for essential medications and medical equipment could be disrupted for months or even years. Preparing in advance with a comprehensive medical kit ensures that you have the resources necessary to treat injuries, manage illnesses, and address long-term health needs.

Here are the key reasons medical preparation is essential:

Injuries from the blast: Survivors of a nuclear explosion may suffer from burns, cuts, fractures, and other traumatic injuries caused by the blast, falling debris, or fires. Having the proper medical supplies can help stabilize and treat these injuries in the critical hours and days after the event.

Radiation sickness: Radiation exposure can lead to acute radiation syndrome (ARS), which requires prompt treatment to reduce the risk of fatality. Stocking medications and treatments for radiation exposure is critical for survival in a fallout zone.

Infection and illness: Cuts, burns, and other injuries are at high risk of infection, especially in a post-disaster environment where hygiene may be compromised. In addition, the breakdown of sanitation systems could lead to outbreaks of diseases like dysentery, cholera, and respiratory infections.

Chronic health conditions: People with chronic health conditions, such as diabetes, high blood pressure, asthma, or heart disease, will need to ensure they have an adequate supply of medications and treatment options to manage their condition over the long term.

Mental health: The psychological toll of surviving a nuclear event and living in a fallout zone can lead to anxiety, depression, and post-traumatic stress disorder (PTSD). Mental health care is just as important as physical care in a survival scenario.

Essential Medical Supplies for Survival

A well-stocked medical kit should contain supplies for treating a wide range of injuries and illnesses, including trauma care, infection prevention, chronic disease management, and radiation exposure. Below is a list of essential medical supplies that should be part of your survival stockpile.

Trauma Care Supplies

Traumatic injuries, such as burns, cuts, fractures, and severe bleeding, are common after a nuclear event. Having the right supplies to stop bleeding, stabilize broken bones, and treat burns is essential.

First aid manual: A comprehensive first aid manual is invaluable, especially if you are not familiar with medical procedures. Make sure the manual covers trauma care, wound care, and other survival-related medical issues.

Bandages and gauze: Stock various sizes of bandages, including adhesive bandages (Band-Aids), sterile gauze pads, and rolls of gauze to wrap wounds. Elastic bandages are also useful for supporting sprains or holding dressings in place.

Tourniquet: In cases of severe bleeding, a tourniquet may be necessary to stop blood loss until proper medical care can be administered. Ensure you have a proper tourniquet and know how to use it safely.

Sterile gloves: Disposable sterile gloves are essential for protecting yourself and the patient from infection when treating open wounds or handling bodily fluids.

Burn dressings and gels: Burn injuries are common after explosions. Stock specialized burn dressings, burn ointments, or hydrogel dressings to treat burns and reduce pain.

Sutures or wound closure strips: For deep cuts and lacerations that need to be closed, having sutures or wound closure strips (butterfly bandages or Steri-Strips) is important. If you're unfamiliar with suturing, learn basic wound closure techniques or opt for adhesive strips as an alternative.

Splints and slings: If someone suffers a fracture or sprain, having a splint and sling can help immobilize the injured limb and reduce pain. Make sure you know how to apply a splint correctly.

Antiseptic solutions and wipes: Stock antiseptic wipes, hydrogen peroxide, or iodine solutions to clean wounds and prevent infection.

Medications

Access to medications will likely be limited after a nuclear disaster, so it's essential to stock up on a range of over-the-counter (OTC) and prescription medications. In addition to medications for treating specific conditions, stock a variety of general medications to address common health issues.

Pain relievers: OTC pain relievers like ibuprofen, acetaminophen, and aspirin are essential for managing pain, reducing fever, and addressing inflammation.

Antibiotics: Bacterial infections will be common after injuries or exposure to unsanitary conditions. If possible, stock antibiotics such as amoxicillin, ciprofloxacin, or doxycycline. Be aware of proper dosages and indications for use, and consult a healthcare provider if possible before stockpiling prescription medications.

Antidiarrheal medications: Diarrhea caused by contaminated food or water can lead to dehydration and serious complications. Stock medications like loperamide (Imodium) to treat diarrhea.

Anti-nausea medications: Radiation exposure and infections can cause nausea and vomiting. Stock anti-nausea medications such as meclizine or ondansetron to manage these symptoms.

Cold and flu medications: Prepare for respiratory infections with medications like decongestants, antihistamines, and cough suppressants.

Potassium iodide (KI): Potassium iodide is a crucial medication for protecting the thyroid from radioactive iodine exposure. Ensure you have enough KI tablets for everyone in your group and understand the appropriate dosages based on age and exposure levels.

Allergy medications: Antihistamines like diphenhydramine (Benadryl) can help manage allergic reactions and insect bites.

Antacids and stomach medications: Stock antacids like calcium carbonate (Tums) and medications for indigestion or acid reflux, as stress and poor nutrition can exacerbate these issues.

Chronic condition medications: If you or someone in your group has a chronic condition like diabetes, high blood pressure, asthma, or heart disease, ensure you have an adequate supply of prescription medications. Work with your healthcare provider to stock extra medications if possible, and consider alternatives like insulin alternatives or herbal remedies in case supplies run out.

Radiation Exposure Supplies

In a fallout zone, radiation exposure is one of the primary health concerns. Stock supplies to mitigate radiation exposure and monitor radiation levels.

Radiation detectors (Geiger counters): A Geiger counter is essential for measuring radiation levels in the environment, food, and water. Ensure you have a reliable device and know how to use it.

Dosimeters: Personal dosimeters track cumulative radiation exposure over time. These are useful for monitoring how much radiation each person has been exposed to and determining when it's necessary to evacuate or seek additional shelter.

Potassium iodide (KI): As mentioned, KI tablets are essential for protecting the thyroid from absorbing radioactive iodine. Stock enough for everyone in your group, and be familiar with the recommended dosages for children and adults.

Protective clothing: Stock full-body protective suits, masks, and gloves to reduce exposure when going outside in contaminated areas. N95 or P100 masks can help reduce the risk of inhaling radioactive particles.

Infection Control and Hygiene Supplies

Maintaining hygiene in a survival scenario is critical for preventing infections and the spread of disease. Stock essential hygiene and infection control supplies.

Hand sanitizer: Stock alcohol-based hand sanitizer for cleaning hands when water and soap are not available.

Soap and disinfectants: Ensure you have a supply of antibacterial soap and disinfectants like bleach to clean surfaces and maintain hygiene.

Masks and gloves: Surgical masks and disposable gloves can help reduce the risk of infection, especially when treating injuries or caring for someone who is sick.

Water purification tablets: Ensure your water supply is safe by stockpiling water purification tablets or portable filters to remove harmful bacteria, viruses, and contaminants from water.

Chronic Condition and Long-Term Care Supplies

For those with chronic health conditions, such as diabetes, asthma, or heart disease, having the right supplies and medications is crucial for long-term survival.

Insulin and syringes: If you or someone in your group has diabetes, stock up on insulin and syringes. Learn how to properly store insulin (e.g., in a cool, dry place) and consider alternatives like insulin-stimulating herbs if your supply runs out.

Blood pressure monitors: For people with hypertension or heart conditions, having a blood pressure monitor is essential for tracking health and managing medications effectively.

Asthma inhalers and spacers: Stock up on asthma inhalers, spacers, and medications to manage respiratory conditions. Keep an extra supply of bronchodilators and corticosteroids if possible.

Mental Health Support Supplies

Surviving a nuclear event and living in a fallout zone can take a significant toll on mental health. Preparing for mental health care is just as important as physical health in a survival situation.

Anti-anxiety medications: Stock medications like benzodiazepines (if prescribed) or over-the-counter supplements like valerian root or melatonin for managing anxiety and insomnia.

Antidepressants: If you or someone in your group takes antidepressants, try to stockpile enough to last for several months or more. Work with a healthcare provider to ensure a stable supply.

Stress-relief tools: Consider stocking stress-relief tools like journals, books, or stress balls. These can help reduce anxiety and provide a sense of normalcy during difficult times.

Basic Medical Tools

Having the right tools to provide basic medical care is essential for treating injuries and managing health in a survival scenario.

Thermometer: A reliable thermometer is necessary for monitoring body temperature, particularly when dealing with infections or fever.

Tweezers and scissors: Stock medical-grade tweezers for removing splinters or debris from wounds and scissors for cutting bandages and clothing.

Syringes and needles: Stock sterile syringes and needles for administering medications, drawing blood, or treating wounds.

Stethoscope and blood pressure cuff: These tools can help you monitor vital signs and detect issues like low blood pressure or respiratory problems.

Developing Medical Skills for Survival

In addition to stocking medical supplies, it's essential to develop basic medical skills to provide care in the absence of professional medical help. Consider taking a first aid and CPR course, learning wound care techniques, and studying common medical conditions and treatments. Having a basic understanding of trauma care, infection control, and chronic disease management will improve your ability to handle medical emergencies in a survival situation.

Wound care: Learn how to properly clean, dress, and close wounds to prevent infection.

CPR and resuscitation: Knowing how to perform CPR and basic life support can save lives in critical situations.

Managing chronic conditions: Familiarize yourself with the symptoms, treatments, and management of chronic conditions like diabetes, asthma, and heart disease.

Medical preparation is a crucial aspect of surviving in a post-nuclear world. By stocking up on essential medical supplies, medications, and tools, and developing basic medical skills, you can increase your chances of surviving injuries, illnesses, and long-term health challenges. A well-prepared medical kit not only helps you manage immediate medical needs but also ensures that you can address long-term health issues in the absence of professional medical care.

Maintaining a Power Source in the Post-Nuclear World

After a nuclear disaster, the modern infrastructure that powers our homes, hospitals, and communication networks would likely be severely damaged or entirely destroyed. In the absence of a functioning electrical grid, maintaining a reliable power source becomes essential for survival, enabling you to cook food, power medical equipment, charge communication devices, and provide heat during cold conditions. In this chapter, we'll explore the challenges of maintaining a power source in the post-nuclear world, the best alternative energy options, and practical strategies for generating and conserving electricity.

Why Power is Essential for Survival

In a post-nuclear environment, power is not just a convenience—it is a critical resource for survival. Access to electricity enables you to do the following:

Communication: Powering radios, shortwave radios, and other communication devices is essential for staying informed about radiation levels, weather conditions, and rescue efforts.

Lighting: Light sources are necessary for safety, particularly when the sun may be obscured by the atmospheric effects of a nuclear winter. Artificial lighting can also help maintain a sense of normalcy and reduce the psychological toll of isolation.

Cooking and heating: A reliable power source allows you to cook food and boil water, which is vital in areas where fuel may be scarce or contaminated. Heating is also essential in cold environments or during nuclear winter when temperatures may drop significantly.

Medical needs: Power is crucial for operating medical devices, such as refrigeration for medications like insulin or running life-saving equipment.

Water purification: Electric water purification systems can remove contaminants from water, ensuring a safe drinking supply in areas affected by radioactive fallout.

Given the importance of electricity in a survival scenario, finding alternative energy sources that are sustainable and adaptable to a post-nuclear world is a top priority.

Challenges of Power Generation in a Post-Nuclear World

Several factors make power generation in a post-nuclear environment particularly challenging:

Damaged infrastructure: Power plants, substations, and transmission lines may be destroyed or rendered inoperable by nuclear explosions or the electromagnetic pulse (EMP) generated by the blasts. This could lead to long-term outages across entire regions or countries.

Limited access to fuel: Traditional power sources like coal, oil, and natural gas may become difficult to access, either because of supply chain disruptions or contamination. Transportation of fuel may also be hindered by damaged roads and infrastructure.

Environmental conditions: The effects of nuclear winter—reduced sunlight, extreme cold, and unpredictable weather patterns—could hinder solar power generation and make outdoor work difficult or dangerous.

Alternative Power Sources for Survival

To maintain a reliable power source after a nuclear disaster, you'll need to turn to alternative energy solutions. Below are some of the best options for generating electricity in a post-nuclear environment, along with their advantages and challenges.

Solar Power

Solar power is one of the most accessible and sustainable energy sources for survival. With the right setup, you can generate electricity from sunlight, providing power for your essential devices and needs.

Advantages:

Solar panels are relatively easy to install, portable, and scalable.

They generate power silently and do not rely on fuel, making them a long-term solution for power generation.

Solar energy can be stored in batteries, providing electricity even when the sun isn't shining.

Challenges:

Solar panels rely on sunlight, which could be limited during a nuclear winter or overcast conditions caused by fallout clouds.

Initial setup costs can be high, and panels must be kept clean and free from debris to function effectively.

Solar panels are vulnerable to damage from high winds or debris in post-disaster environments.

Best uses: Solar power is ideal for charging small devices like radios, cell phones, and LED lights. With a larger setup, you can power refrigerators, medical equipment, and water filtration systems.

Wind Power

Wind turbines harness the energy of the wind to generate electricity. Small-scale wind turbines are available for home use and can be an effective way to produce power in areas with consistent wind.

Advantages:

Wind power is renewable and does not require fuel, making it a long-term solution.

Wind turbines can generate electricity even when the sun isn't shining, providing a backup to solar power.

Challenges:

Wind turbines are dependent on wind availability, and wind patterns may be disrupted by changes in weather caused by nuclear winter.

Setting up wind turbines requires space, and they are vulnerable to damage from debris or high winds.

Noise from wind turbines may be a concern, especially in areas where quiet is essential for security.

Best uses: Wind power is best suited for locations with consistent wind, such as open plains or coastal areas. It can be used to supplement solar power and charge batteries for later use.

Hydropower

Hydropower uses flowing water to generate electricity. In areas with access to rivers or streams, small-scale hydropower systems can provide a steady and reliable source of power.

Advantages:

Hydropower systems can generate electricity continuously as long as water is flowing, making it one of the most reliable renewable energy sources.

It provides a consistent power output, unlike solar and wind, which are weather-dependent.

Challenges:

Access to water is essential, and in a fallout scenario, water sources may be contaminated with radiation.

Building a hydropower system requires technical knowledge and the ability to install turbines in a flowing water source.

Droughts or changes in water levels (which may occur due to nuclear winter) could limit power generation.

Best uses: Hydropower is ideal for areas with rivers, streams, or waterfalls. It is well-suited for continuous power generation, providing electricity for homes, lighting, and small appliances.

Hand-Crank Generators

Hand-crank generators allow you to generate small amounts of electricity by manually turning a crank. These generators are highly portable and do not rely on external energy sources like fuel, sunlight, or wind.

Advantages:

Hand-crank generators are compact, portable, and can be used indoors.

They provide a reliable source of power for small devices like radios, flashlights, and cell phones.

They do not rely on weather conditions or fuel supplies, making them ideal for emergency situations.

Challenges:

Hand-crank generators produce only small amounts of electricity, which limits their use to low-power devices.

They require physical effort to operate, and extended use can be tiring.

Best uses: Hand-crank generators are best for charging emergency communication devices, small electronics, and lighting during short-term power outages.

Bicycle Generators

Bicycle generators work by converting the kinetic energy of pedalling into electricity. These systems can be built using a stationary bike and a generator, providing a low-tech but effective way to generate power.

Advantages:

Bicycle generators are simple to build and use, requiring only human effort to generate power.

They provide exercise and a sense of activity, which can help alleviate boredom and maintain fitness in a survival scenario.

Challenges:

Like hand-crank generators, bicycle generators produce small amounts of electricity, limiting their use to low-power devices.

Physical effort is required, and extended use can be tiring.

Best uses: Bicycle generators are useful for charging small devices, powering lights, and providing an emergency power source when other options are unavailable.

Battery Banks

Battery banks store electricity generated by solar panels, wind turbines, or other renewable energy sources for later use. Having a reliable battery storage system allows you to collect power during the day and use it when needed, even if conditions aren't favorable for power generation.

Advantages:

Battery banks allow you to store energy for use during the night or when power generation is low.

They provide a stable and continuous power supply, even when renewable energy sources fluctuate.

Challenges:

Battery systems can be expensive, and their lifespan is limited. Over time, batteries degrade and need to be replaced.

They require regular maintenance and should be stored in a dry, temperature-controlled environment to prevent damage.

Best uses: Battery banks are ideal for providing backup power during times when renewable energy sources are not available. They can power lights, small appliances, and essential medical equipment.

Conserving Power: Maximizing Efficiency in a Post-Nuclear World

In a post-nuclear world, power generation will be limited, so conserving electricity is just as important as generating it. Here are strategies to maximize energy efficiency and make the most of your power supply:

Use Energy-Efficient Devices

Switch to energy-efficient devices, such as LED lights, low-power radios, and efficient appliances. LED lights, for example, use significantly less energy than traditional incandescent bulbs and last longer.

Charge Devices Sparingly

Charge electronic devices only when necessary. Turn off devices when they're not in use to conserve battery life. Consider using manual alternatives for lighting and communication, such as hand-crank radios or solar-powered flashlights.

Limit Heating and Cooling

Heating and cooling consume large amounts of energy, so conserve power by insulating your shelter with blankets, plastic sheeting, or other materials to trap heat. Wear layers of clothing to stay warm, and use portable heaters sparingly.

Utilize Natural Light

During daylight hours, maximize the use of natural light by positioning work areas near windows or using reflective surfaces to brighten your space. This reduces the need for electric lighting.

Plan Power Usage

Prioritize essential power needs, such as medical equipment, lighting, and communication devices. Avoid wasting power on unnecessary tasks or entertainment, and plan your energy usage carefully to ensure you have enough power when it's needed most.

Practical Steps for Setting Up an Off-Grid Power System

If you're preparing to maintain power in a post-nuclear world, here are the basic steps to setting up an off-grid power system:

Assess your power needs: Determine how much electricity you'll need based on the devices and equipment you plan to power. Calculate the wattage requirements of each device and estimate how many hours per day they'll be used.

Choose your power sources: Based on your environment and available resources, decide which power sources will work best for your situation. Solar panels, wind turbines, and battery banks are commonly used in off-grid systems.

Install renewable energy systems: Set up solar panels, wind turbines, or hydropower systems to generate electricity. Position solar panels in areas that receive maximum sunlight and ensure that wind turbines are installed in open, windy areas.

Set up a battery bank: Install a battery storage system to store excess electricity generated by your renewable energy systems. This ensures that you have power available even when generation is low.

Connect your devices: Use an inverter to convert the direct current (DC) power generated by your renewable systems into alternating current (AC) power, which most household devices use. Connect your essential devices to the power system.

By exploring alternative energy sources like solar, wind, and hydropower, and by setting up an off-grid system, you can ensure a reliable and sustainable power supply even in the most challenging conditions. Combining energy conservation strategies with practical power generation solutions will help you maximize your resources and maintain a functional, resilient lifestyle in a world where the electrical grid may no longer exist.

Solar Energy and Fallout: Can it Still Work?

In the aftermath of a nuclear event, many survivors may turn to solar energy as a potential power source, especially given its sustainability and low maintenance. However, the effectiveness of solar energy in a post-nuclear environment raises important questions, especially concerning fallout, radiation, and the possible effects of a nuclear winter. This chapter will explore whether solar energy can still work after a nuclear disaster, how fallout and nuclear winter could impact solar power generation, and strategies for maximizing the efficiency of solar panels in these conditions.

Solar Energy Basics: How Solar Panels Work

Solar panels generate electricity by converting sunlight into direct current (DC) electricity using photovoltaic (PV) cells. These cells are made from semiconductor materials, typically silicon, that generate electricity when they are exposed to sunlight. The more sunlight the panels receive, the more electricity they produce. Solar panels are often coupled with battery storage systems, allowing users to store excess electricity for later use.

Solar power is a popular choice for off-grid energy solutions because it is renewable, scalable, and silent. Once installed, solar panels require minimal maintenance and can provide power for years without relying on external fuel sources.

Fallout and Its Impact on Solar Panels

After a nuclear explosion, radioactive fallout consisting of dust, soot, and radioactive particles will settle on the ground and on any exposed surfaces, including solar panels. Fallout can impact solar energy production in several ways:

Blocking Sunlight

One of the most significant concerns is the effect of fallout particles settling on the surface of solar panels, blocking sunlight and reducing their efficiency. Dust and soot can accumulate quickly after a nuclear event, particularly in the immediate fallout zone, and this can reduce the amount of sunlight that reaches the photovoltaic cells.

Impact on efficiency: Even a thin layer of dust or ash can significantly reduce solar panel efficiency. Studies show that a 5% to 10% reduction in solar power generation can occur from minor soiling, while heavy soiling from ash or fallout could reduce efficiency by 20% to 30% or more.

Radiation and Photovoltaic Cells

Another concern is whether radiation could damage the photovoltaic cells in solar panels. Fortunately, most solar panels are designed to be robust and resistant to environmental damage, including UV radiation from the sun. The types of radiation released during a nuclear event—such as alpha, beta, and gamma radiation—are unlikely to damage the silicon-based photovoltaic cells directly.

Radiation impact: While radiation is a danger to living organisms, it is unlikely to degrade the materials in solar panels significantly. However, long-term exposure to high levels of radiation could affect other components of a solar system, such as wiring or electronics in inverters and charge controllers.

Temperature and Weathering Effects

Solar panels are generally designed to withstand a range of temperatures, but a nuclear winter scenario could bring extreme weather changes. In a nuclear winter, the atmosphere could cool significantly, and weather patterns may become erratic, with longer winters, colder temperatures, and possible changes in wind and precipitation patterns. This could affect solar power generation in several ways:

Cold temperatures: Solar panels tend to work more efficiently in cooler temperatures, so while colder conditions might not inhibit electricity generation, reduced sunlight exposure during a nuclear winter could lower overall power output.

Snow and ice: Snow or ice accumulation on solar panels during a nuclear winter could further block sunlight. Regular clearing of panels may be necessary to maintain power generation.

Nuclear Winter and Its Impact on Solar Power

A nuclear winter is a potential climatic aftermath of a large-scale nuclear conflict. It occurs when massive amounts of soot, ash, and smoke from firestorms rise into the upper atmosphere, blocking sunlight and dramatically lowering global temperatures. The resulting "winter" could last for months or even years, depending on the scale of the nuclear event.

Reduced Sunlight

The primary issue with nuclear winter for solar power generation is the reduced sunlight. With soot and ash particles suspended in the stratosphere, much of the sun's light and heat could be blocked, leading to significantly dimmer conditions on the ground.

Sunlight reduction: Some studies estimate that nuclear winter could result in a reduction of sunlight by 50% to 70% or more, depending on the severity of the event. In this scenario, solar panels would still generate electricity, but the output would be dramatically reduced.

Cloudy conditions: Solar panels can still generate electricity in cloudy conditions, but at a reduced rate. On overcast days, panels typically produce 10% to 25% of their maximum output. In a nuclear winter scenario, this reduced output could persist for an extended period, making solar energy less reliable as a primary power source.

Longer Nights and Shorter Days

In addition to reduced sunlight, nuclear winter could result in longer nights and shorter days, further limiting the time that solar panels can generate electricity each day. This means that energy storage solutions, such as battery banks, become even more critical for maintaining power during the extended night periods.

Maximizing Solar Energy Efficiency in Fallout Zones

Despite the challenges posed by fallout and nuclear winter, solar power can still be a valuable energy source for survivors. By taking proactive steps to maximize solar panel efficiency and adapt to changing conditions, you can ensure that your solar system provides at least some power during and after a nuclear disaster.

Regular Cleaning and Maintenance

One of the most important steps in maintaining solar energy efficiency in a fallout zone is regularly cleaning your solar panels. Fallout dust and ash will quickly accumulate on the panels, reducing their ability to capture sunlight.

Cleaning frequency: Clean your solar panels frequently—daily if necessary—using non-abrasive materials and water (preferably filtered or purified to avoid contamination). If water is scarce, a soft brush or cloth can be used to remove most of the debris.

Safety precautions: When cleaning panels in a fallout zone, always wear protective gear, including gloves, a mask, and long sleeves, to minimize exposure to radioactive particles. Store contaminated cleaning materials in sealed containers to prevent further spread of fallout.

Optimize Panel Placement

The placement of your solar panels can have a significant impact on their ability to generate electricity. Even in reduced-light conditions, proper positioning can help maximize the amount of sunlight your panels receive.

Maximize sunlight: Ensure your panels are positioned to capture the maximum amount of sunlight each day. In the northern hemisphere, this typically means facing the panels southward at an angle optimized for your latitude. Adjust the angle seasonally to account for the sun's changing position.

Remove obstructions: If possible, clear away any obstructions like debris, trees, or damaged structures that could cast additional shadows on your panels.

Use Battery Storage

In a post-nuclear world, battery storage systems become even more critical. Given the reduced amount of sunlight during a nuclear winter or overcast conditions, solar panels may only generate electricity for a few hours a day. By storing excess electricity in batteries, you can ensure that power is available when sunlight is not.

Battery capacity: Invest in high-capacity battery storage systems capable of storing enough electricity to last through extended periods of low sunlight. Lithium-ion batteries are commonly used for solar energy storage due to their long lifespan and efficiency.

Energy conservation: Conserve electricity by using it sparingly and only for essential devices and tasks. Prioritize communication devices, medical equipment, and lighting, and avoid unnecessary power consumption.

Combine Solar with Other Power Sources

To increase your chances of maintaining a reliable power supply, consider combining solar energy with other alternative power sources, such as wind turbines or hand-crank generators. Diversifying your power generation methods can help ensure that you have electricity even when solar energy production is low.

Wind turbines: Wind power can complement solar energy, especially in areas where wind is consistent. Wind turbines can generate electricity at night and during cloudy conditions, providing a backup to solar power.

Hand-crank generators: In emergencies, hand-crank generators can be used to power small devices like radios or lights. They require no fuel or external energy source, making them a valuable backup option.

Energy-Efficient Devices

To make the most of the limited power you generate, switch to energy-efficient devices. LED lights, for example, use significantly less energy than incandescent bulbs, and low-power appliances can help stretch your stored electricity.

LED lights: Use LED lighting in place of traditional bulbs to reduce energy consumption while still providing bright illumination.

Low-power radios and communication devices: Choose energy-efficient radios, shortwave transmitters, and other essential communication tools that require minimal power.

Solar energy can still work in the aftermath of a nuclear event, but its effectiveness will be limited by fallout, nuclear winter conditions, and the availability of sunlight. While solar panels may produce less electricity due to reduced sunlight and fallout debris, they remain a viable option for powering essential devices, particularly when paired with battery storage and regular maintenance. By cleaning panels regularly, optimizing their placement, and combining solar power with other energy sources, you can continue to generate electricity even in the challenging conditions of a post-nuclear world. Solar energy, while not perfect in these conditions, can still be a valuable tool for survival when properly managed and supplemented with other power generation methods.

Bartering in the Nuclear Winter: How to Trade Goods and Skills

In a post-nuclear world, where modern economies and supply chains have collapsed, bartering will become a primary method of survival. With currency likely rendered useless due to hyperinflation or lack of trust in its value, people will turn to the direct exchange of goods and services to meet their needs. The importance of bartering cannot be overstated in such a scenario, as it provides a means to acquire essential supplies, tools, and skills necessary for survival in a harsh, resource-scarce environment. This chapter will explore how bartering will work during a nuclear winter, what goods and skills will be most valuable, and practical strategies for successful trading.

Why Bartering Becomes Essential in a Post-Nuclear World

In the aftermath of a nuclear disaster, the entire global economy will likely be disrupted. Financial systems may collapse, banking may become inaccessible, and traditional markets may no longer exist. Trade routes may be blocked, cities destroyed, and infrastructure obliterated. In this context, people will need to rely on local resources and immediate communities to survive, and bartering will emerge as the most practical means of exchanging value.

Bartering allows survivors to trade essential goods and services that may no longer be obtainable through conventional means. In a world where manufactured goods are scarce, and production has ground to a halt, bartering becomes the only reliable way to get what you need—whether it's food, water, medical supplies, or knowledge.

Key Goods for Bartering in a Nuclear Winter

Certain goods will become incredibly valuable in a post-nuclear environment, particularly those that are essential for survival or difficult to produce. Understanding which items will be in demand and having the foresight to stockpile or secure these items can give you a significant advantage in bartering. Here are some of the most valuable goods to have for bartering during a nuclear winter:

Food and Water

In a nuclear winter, food and water are the most critical resources for survival, making them top-tier barter items. Clean, uncontaminated water will be hard to come by in fallout zones, and food production will be severely disrupted by reduced sunlight, contaminated soil, and the breakdown of agricultural systems.

Dried and canned food: Non-perishable food items, such as dried beans, rice, pasta, and canned goods, will be incredibly valuable. These foods are easy to store, have a long shelf life, and provide essential nutrients.

Water purification: Water will likely be contaminated by fallout, making water purification systems or tablets extremely valuable. If you have the ability to purify water, you can trade purified water or water filters for other essentials.

Medical Supplies

With limited access to medical professionals and hospitals, medical supplies will become vital for treating injuries, illnesses, and radiation exposure. Bartering medical items can make the difference between life and death in many situations.

First aid supplies: Bandages, gauze, antiseptics, pain relievers, and burn treatments will be in high demand. These items can be traded for food, tools, or other necessities.

Medications: Antibiotics, anti-diarrheal medications, and potassium iodide (KI) tablets to prevent radioactive iodine absorption will be among the most sought-after items. People will trade valuable resources for medicine, especially if they or their loved ones are sick.

Sanitary items: Sanitary products like soap, hand sanitizer, and disinfectants will be critical for hygiene and preventing disease, making them valuable barter goods.

Fuel and Energy Sources

Fuel for heating, cooking, and power generation will be a rare commodity. Those with access to fuel or alternative energy sources can trade these items for nearly anything they need.

Firewood: In colder regions, firewood or any material that can be used to generate heat will be extremely valuable, especially during the nuclear winter. Those who can supply wood or fuel will have significant bartering power.

Propane and gasoline: Propane for cooking and gasoline for generators or transportation will also be in high demand. Storing small amounts of these fuels can give you a valuable bartering tool.

Batteries and solar chargers: Rechargeable batteries, solar-powered lights, and hand-crank generators will also be useful as people struggle to maintain power for essential devices like radios and flashlights.

Tools and Survival Gear

Tools that can help people rebuild, repair, or create will be essential in a post-nuclear world. Additionally, survival gear that helps people stay safe in the harsh environment of a nuclear winter will hold significant value.

Hand tools: Basic tools like hammers, wrenches, knives, screwdrivers, and saws will be highly valuable for repairing shelters, creating new structures, or performing basic survival tasks.

Hunting and fishing gear: If food production has collapsed, hunting and fishing may become essential for survival. Items like fishing lines, hooks, traps, or hunting knives can be valuable for those needing to secure their own food.

Fire-starting materials: Lighters, matches, or fire-starting kits will be essential for creating heat and cooking. Stockpiling these items can make you a valuable bartering partner.

Warm clothing and blankets: In the cold conditions of a nuclear winter, warm clothing, blankets, and insulation materials will be in high demand. People will trade food and tools for the means to stay warm.

Seeds and Gardening Supplies

While agriculture will be disrupted by nuclear fallout, seeds and gardening tools will become incredibly valuable as survivors look for ways to produce their own food. Heirloom seeds that can be saved and replanted year after year will be particularly sought after.

Non-GMO seeds: Seeds for hardy crops like potatoes, carrots, beans, and cabbage that can grow in poor soil conditions or hydroponically will be valuable for people trying to grow food in difficult conditions.

Gardening tools: Shovels, hoes, watering cans, and compost can also be traded as people attempt to grow their own food indoors or in protected outdoor areas.

Clothing and Footwear

In a world where manufacturing has collapsed, clothing and footwear will become scarce. Durable clothing that can withstand harsh conditions will be particularly valuable.

Sturdy boots: Footwear that can last for long treks or difficult terrain will be highly prized. People will trade essential items for boots that protect them from the elements.

Durable clothing: Clothing made from tough materials, like wool or heavy-duty fabrics, will also be in demand, especially as people look for warmth and protection from the fallout environment.

Communication Devices

In the absence of a functioning electrical grid and traditional communication systems, radios and other forms of communication will be critical. Information will be one of the most valuable commodities in a post-nuclear world.

Radios: Hand-crank or battery-powered radios will be vital for staying connected with the outside world, listening for news, and coordinating with other survivors. Radios will be traded for high-value goods such as food, medicine, or weapons.

Walkie-talkies: Short-range communication devices like walkie-talkies will be important for coordinating within a community or family, and they can be traded for other essential supplies.

Bartering Skills for Survival

In addition to physical goods, skills will be equally valuable in a post-nuclear world. People who possess knowledge and skills that others lack can trade their abilities for food, shelter, or other necessities. Here are some of the most valuable skills for bartering:

Medical Skills

If you have medical training, your services will be in extremely high demand. Bartering medical care, including wound treatment, setting broken bones, and administering medications, can ensure you receive a steady supply of food, water, or other essentials.

Mechanical and Repair Skills

The ability to repair tools, vehicles, or generators will be valuable in a world where new equipment is hard to come by. Mechanics, electricians, and those who can fix engines or solar systems will have an essential skill to trade.

Agricultural Skills

If you know how to grow food in difficult conditions, such as indoor gardening, hydroponics, or growing in contaminated soil, you can trade your expertise for supplies. Teaching others how to grow food or manage livestock will be an invaluable service.

Construction and Carpentry

Building or repairing shelters, fortifications, or makeshift infrastructure will be crucial. People with carpentry skills, construction knowledge, or engineering expertise will find plenty of opportunities to trade their labor for essential goods.

Weapon Repair and Maintenance

In a post-nuclear world, weapons will be important for defense, hunting, and maintaining order. If you have skills in weapon repair or maintenance, such as fixing firearms or crafting makeshift weapons, you can barter your expertise for critical supplies.

Sewing and Clothing Repair

The ability to repair clothing, sew new garments, or patch up worn-out materials will be valuable. People will trade food, tools, and other goods for someone who can extend the life of their clothing in a world where new clothing is scarce.

Practical Bartering Strategies

Bartering in a post-nuclear world requires not only understanding what is valuable but also knowing how to engage in successful trades. Here are some practical strategies to help you barter effectively:

Establish Trust

Trust will be a key component of successful bartering, especially in a survival scenario where people are desperate. Building a reputation as someone who offers fair trades and can be relied upon is essential.

Honesty: Be transparent about the quality of the goods or services you are offering. Misleading people will harm your reputation and could lead to conflict.

Reliability: If you promise to deliver something, make sure you follow through. Consistency builds long-term relationships and encourages repeat trades.

Understand the Value of Goods

The value of goods will change depending on the availability and need in your region. Certain items that seem common now—like fresh water, food, or medical supplies—may become incredibly scarce and valuable.

Supply and demand: Be aware of what is in short supply in your community, and adjust your bartering strategy accordingly. For example, water may be more valuable in an area suffering from fallout contamination, while firewood may be prized in a colder region.

Stockpile essentials: Before disaster strikes, stockpile items that are likely to be in high demand. Anticipating shortages can give you the upper hand in future bartering scenarios.

Barter in Groups or Communities

Bartering within a group or community can offer greater security and allow for more effective exchanges. Working as part of a group, such as a survival community or local trade network, allows for better resource management and collective bargaining power.

Local trade networks: Establishing a local trade network can help create a stable barter economy. This can involve organizing regular barter markets or exchanges where community members gather to trade goods and services.

Security: Trading in groups also provides protection against theft or exploitation. Always be cautious when bartering with strangers or unknown groups, especially if the trade involves valuable items.

Be Adaptable

The value of certain goods and skills may fluctuate over time. Be prepared to adapt your bartering strategy as conditions change.

Flexibility: Don't become overly reliant on a single item for bartering. Diversify the goods and skills you offer to ensure you can continue trading even if the demand for one item decreases.

Long-term trading relationships: Building strong, long-term relationships with others in your community can help ensure a steady flow of trades, even as market conditions change.

Bartering will become a vital means of survival in the wake of a nuclear winter, allowing people to trade essential goods and skills in a world where traditional currencies and economies have collapsed. By understanding what items and services will be most valuable, preparing a stockpile of key goods, and developing critical survival skills, you can position yourself as a successful barterer in this new, harsh reality. Establishing trust, being adaptable, and working within your community will help you navigate the challenges of post-nuclear bartering, ensuring you have access to the resources needed to survive and thrive in a world forever changed by nuclear disaster.

How to Communicate in a World Without Technology

In a post-nuclear world, the collapse of modern infrastructure would likely mean the loss of communication technologies we take for granted—cell phones, internet, and even traditional landlines could become useless. Electromagnetic pulses (EMPs) from nuclear explosions could knock out electronic devices, and the destruction of power grids and telecommunications infrastructure would sever most means of instant communication. However, effective communication is still essential for survival, coordination, and safety. This chapter will explore alternative communication methods in a world without modern technology, focusing on low-tech and no-tech solutions for staying connected with others.

Why Communication is Critical for Survival

In the wake of a nuclear disaster, the ability to communicate with others becomes crucial for several reasons:

Coordination of resources: Sharing information about available food, water, medical supplies, and shelter can help communities survive and make efficient use of scarce resources.

Security and safety: Communication allows survivors to warn others of dangers, including radioactive zones, hostile groups, or environmental threats. It's also essential for coordinating defense and safety efforts.

Mental well-being: Isolation can take a psychological toll. Being able to communicate with others provides emotional support, fosters cooperation, and reduces the stress of surviving in harsh conditions.

Trade and bartering: Maintaining communication channels is necessary for establishing trade networks and bartering goods and services, which are crucial for long-term survival.

The Impact of EMPs on Communication

One of the most significant threats to modern technology in a post-nuclear world is the potential for electromagnetic pulses (EMPs) to disable electronic devices. An EMP is a burst of electromagnetic energy that can be caused by nuclear explosions, particularly at high altitudes. EMPs have the power to:

Fry electrical circuits: EMPs can disable electronic devices, from smartphones and computers to radios and power grids. Unshielded devices may be rendered useless in an instant.

Disrupt communication networks: EMPs can cause widespread outages in communication networks, including cell towers, satellite communications, and traditional broadcasting systems.

While some devices may be hardened or protected against EMPs, most consumer electronics are vulnerable. This means that even if you have a functioning radio or communication device, the networks it depends on may no longer be operational. In such a scenario, low-tech or no-tech communication methods become essential.

Low-Tech Communication Methods

Even without modern technology, several low-tech communication methods can still be highly effective in a post-nuclear world. These methods rely on simple tools, basic electronics, or manual systems that are resilient in a disaster scenario.

Hand-Crank Radios

Hand-crank radios are a critical tool for receiving information in a post-nuclear world. These radios do not require external power sources like batteries or electricity; instead, they are powered by manually turning a crank.

How they work: By cranking the handle, you generate enough power to tune into AM, FM, or shortwave frequencies. Many hand-crank radios also come equipped with built-in flashlights or USB ports for charging small devices.

Use in a post-nuclear world: Hand-crank radios allow you to listen for emergency broadcasts, weather reports, and news about radiation levels or rescue efforts. If EMPs have disabled more advanced communications infrastructure, these radios may still pick up signals from remote or unaffected areas.

Considerations: Stockpile a few hand-crank radios and learn how to tune into shortwave frequencies, which can carry signals over long distances. Be sure to test the radios periodically to ensure they function properly.

Shortwave and Ham Radios

Shortwave and ham radios are two of the most reliable communication methods in a disaster scenario. These radios operate on frequencies that can travel long distances, making them ideal for contacting others outside your immediate area.

Shortwave radios: Shortwave radios can receive signals over vast distances, sometimes spanning continents. Even if local communication systems are down, shortwave signals from unaffected regions may still be available. Hand-crank shortwave radios, in particular, can be useful for receiving broadcasts.

Ham radios: Ham radios (also known as amateur radios) allow two-way communication over a range of frequencies. These radios require a license to operate, but in a survival situation, they may become one of the few ways to communicate over long distances.

Advantages: Ham radios are versatile and can be used to communicate across regions or even countries. With the right equipment, ham radios can transmit voice, text, and even Morse code.

Challenges: Ham radio setups require some technical knowledge to operate. You'll need a working transceiver, an antenna, and a power source (such as a generator, solar power, or a hand-crank power system).

Considerations: It's wise to invest in a ham radio setup and learn how to operate it before disaster strikes. Many preppers take ham radio classes or join local amateur radio clubs to gain practical experience. Ensure that your setup is protected from EMPs by storing it in a Faraday cage (discussed later in this chapter).

Walkie-Talkies

Walkie-talkies are simple two-way radios that allow short-range communication between two or more people. They are portable, easy to use, and do not rely on cell towers or other external infrastructure.

Range: Most walkie-talkies have a limited range, typically between 2 to 5 miles, depending on terrain and obstructions. In open areas with few barriers, the range may be extended, but in urban or mountainous regions, the range will be shorter.

Uses: Walkie-talkies are ideal for communication within small groups or communities. They allow you to stay in contact with family members, neighbors, or fellow survivors over short distances, making them useful for coordinating tasks, organizing defenses, or staying connected during travel.

Considerations: Stockpile walkie-talkies with long battery life or consider models that can be powered by solar chargers or hand-crank generators. Keep extra batteries on hand and ensure your group knows how to use them effectively.

Signal Mirrors and Flares

For visual communication, signal mirrors and flares are highly effective, particularly when trying to signal for help or communicate across distances where verbal communication isn't possible.

Signal mirrors: Signal mirrors reflect sunlight to create bright flashes that can be seen over long distances. The flashes can be used to signal for help or send simple messages, such as "SOS." These mirrors work best in open areas with direct sunlight and can be a lifesaver in remote locations.

Flares: Flares can be used to signal for help, especially at night or in poor visibility. Red flares are commonly associated with distress signals. While flares are single-use, they are an essential part of any survival kit.

Considerations: Keep signal mirrors in your emergency kit, and ensure you know how to use them to aim sunlight effectively. Flares should be kept dry and stored safely to prevent accidental activation.

Morse Code

Morse code is a system of dots and dashes that can be used to communicate over long distances using sound, light, or radio waves. While it may seem outdated, Morse code remains one of the most effective and reliable ways to communicate in low-tech situations.

How it works: Morse code uses a combination of short and long signals (dots and dashes) to represent letters and numbers. These signals can be transmitted using lights (e.g., flashlights or signal mirrors), sound (e.g., tapping on metal, whistling, or using a radio), or radio waves.

Use in a survival situation: If you have access to a ham radio or walkie-talkie, you can use Morse code to transmit messages even when voice communication isn't possible. Morse code is particularly useful in low-signal or noisy environments.

Considerations: Learning Morse code can be a valuable skill in a post-nuclear world. Practice sending and receiving basic messages, such as distress signals (e.g., "SOS") or essential instructions.

No-Tech Communication Methods

In a world where even basic technology may be unreliable or unavailable, no-tech communication methods become vital. These methods rely on human ingenuity and simple tools, ensuring that you can stay connected even without any electronic devices.

Message Boards and Written Notes

When communication devices fail, written messages can serve as a reliable means of communication, particularly within a community or between groups.

Message boards: Establish public message boards in strategic locations, such as community centers, meeting points, or along common travel routes. People can leave notes about available resources, hazards, or requests for help. This creates a central hub for information exchange.

Written notes: In situations where you need to communicate with someone at a distance, handwritten notes or letters can be passed along via runners or dropped off at predetermined locations.

Considerations: Stock up on durable writing materials, such as waterproof paper and pencils. Establish a system for delivering or posting messages within your community.

Whistles and Sound Signals

Simple sound-based communication methods, such as whistles, can be effective for signaling others over short to medium distances. Specific sound patterns can be used to convey different messages, such as warnings or requests for assistance.

Whistles: Whistles can be heard over long distances and through dense environments like forests or urban ruins. Using a specific pattern, such as three short blasts for "help" or one long blast for "all clear," allows for quick, non-verbal communication.

Drums or bells: In larger communities, drums or bells can be used to send signals or gather people for important announcements. Sound-based systems can be established to convey different types of alerts (e.g., danger, meeting, or meal time).

Considerations: Ensure that everyone in your group or community understands the sound signals and what they represent. Practice using these signals regularly to avoid confusion during emergencies.

Runners and Messengers

Before modern technology, runners and messengers were the backbone of communication. In a post-nuclear world, using trusted individuals to carry messages between groups or communities may become a practical solution, particularly when other methods are unavailable.

How it works: Runners carry written or verbal messages between distant locations. In larger communities, designated runners can be tasked with regularly delivering information between groups, allowing for consistent communication even in the absence of technology.

Considerations: Ensure that runners are fit, trustworthy, and familiar with the routes they'll be traveling. Establish safe paths or checkpoints to protect runners from dangers, and provide them with basic supplies for the journey.

Smoke Signals

In open areas, smoke signals can be used as a visual communication method. While less precise than other methods, smoke signals are effective for signaling over long distances.

How it works: A small fire can be built, and blankets or other materials can be used to create controlled bursts of smoke. Different patterns can be used to signal distress, send messages, or gather people.

Considerations: Smoke signals are most effective in clear weather and can be used to signal distant groups or travelers. Be cautious when using fire in dry conditions to avoid unintentional wildfires.

Protecting Communication Devices: Faraday Cages

One way to protect your radios, walkie-talkies, and other electronic devices from EMPs is by storing them in a Faraday cage. A Faraday cage is a container made of conductive materials that shield its contents from electromagnetic pulses.

How it works: The conductive material in a Faraday cage distributes electromagnetic waves around the cage, preventing them from reaching the devices inside. This protects electronic devices from EMPs, ensuring that they remain functional after a nuclear event.

Building a Faraday cage: You can build a simple Faraday cage using a metal container, such as an ammo box or steel trash can. Line the container with non-conductive materials (such as cardboard or foam) to prevent direct contact between the metal and your devices. Place your radios, walkie-talkies, and other critical electronics inside for protection.

Considerations: Test your Faraday cage to ensure it blocks electromagnetic signals by placing a radio or cell phone inside and attempting to use it. If the signal is blocked, the cage is functioning correctly.

In a world without modern technology, communication will be vital for survival, coordination, and safety. By preparing low-tech and no-tech solutions—such as radios, walkie-talkies, signal mirrors, and written messages—you can maintain essential communication channels even in the most challenging conditions. Learning to adapt to these alternative methods and protecting your communication devices from EMPs will help you stay connected with your community, ensure access to critical information, and increase your chances of survival in a post-nuclear world.

Radiation Suits: Are They Necessary for Survival?

Radiation is one of the most dangerous threats in the aftermath of a nuclear disaster, and understanding how to protect yourself from exposure is critical for survival. One of the most common images of protection from radiation is the use of radiation suits, which are often seen in nuclear disaster response scenarios. But the question remains: are radiation suits truly necessary for survival in a post-nuclear environment? This chapter will explore what radiation suits are, how they work, when they are necessary, and what alternative protective measures you can take if you don't have access to one.

What Are Radiation Suits?

Radiation suits, also known as hazmat suits or nuclear protective suits, are specialized garments designed to protect the wearer from radioactive contamination and exposure. These suits are made from materials that block or reduce the penetration of radioactive particles, keeping harmful substances away from the skin and preventing inhalation or ingestion of radioactive dust or fumes.

There are different types of radiation suits, but most fall into two categories:

Particulate Protection Suits

These suits are designed to protect against radioactive dust and particles, but they do not block all types of radiation (especially gamma rays). They are usually made of tightly woven synthetic materials or rubberized fabrics and are intended to keep radioactive particles from coming into contact with the skin.

Examples: These are commonly seen in nuclear cleanup or decontamination efforts. They may include full-body coverage, including gloves, boots, and hoods, as well as respirators or masks to prevent the inhalation of radioactive particles.

Lead-Lined or Heavy-Duty Suits

Lead-lined suits are designed to provide protection against higher levels of radiation, including gamma radiation. These suits are much heavier than particulate protection suits because they incorporate layers of lead or other dense materials that help absorb radiation. However, such suits are not widely used outside of specialized industries like nuclear power plants or research facilities due to their bulk and weight.

Examples: These are typically worn by nuclear workers in high-radiation areas, such as during the maintenance of nuclear reactors or in areas where gamma radiation is present.

How Radiation Affects the Body

Radiation can affect the human body in several ways, depending on the type and intensity of radiation, the duration of exposure, and whether the exposure is external (from the environment) or internal (through inhalation, ingestion, or absorption). The three primary types of radiation you may encounter in a nuclear fallout zone are:

Alpha Radiation

Alpha particles are large and do not penetrate the skin but can be extremely dangerous if inhaled or ingested. Alpha radiation is emitted from radioactive substances like uranium, radon, or plutonium. While a simple protective barrier (like a particulate radiation suit) can block alpha particles, internal exposure can cause serious harm to tissues and organs.

Beta Radiation

Beta particles are smaller and can penetrate the skin but are typically blocked by clothing, glass, or plastic. Like alpha particles, beta particles pose a significant danger if inhaled or ingested, as they can damage internal organs and tissues. Radiation suits can effectively protect against beta radiation by preventing direct contact with the skin and minimizing inhalation risks.

Gamma Radiation

Gamma rays are high-energy electromagnetic waves that can penetrate deeply into the body and damage cells and DNA. Gamma radiation is much more difficult to shield against compared to alpha and beta particles. Thick materials like lead or concrete are required to reduce gamma exposure, and most radiation suits designed for general use are not effective at blocking gamma radiation.

Are Radiation Suits Necessary for Survival?

Radiation suits are not always necessary for survival in a post-nuclear environment, but they can play a critical role in certain high-risk situations. Here's when and why a radiation suit may or may not be essential:

When Radiation Suits Are Necessary

High-Risk Exposure Areas If you are entering or working in an area with high levels of radioactive contamination (such as a nuclear explosion site, reactor meltdown zone, or heavily irradiated area), a radiation suit may be necessary to prevent contamination from radioactive dust and particles. Suits provide a physical barrier that can prevent radioactive substances from contacting your skin or being inhaled.

Decontamination Work Radiation suits are essential for people involved in decontamination efforts, particularly in cleaning up radioactive dust from buildings, vehicles, or clothing. Workers in these situations are at high risk of coming into contact with radioactive particles, making protective suits and respirators crucial.

Evacuation through Fallout Zones If you need to travel through areas heavily affected by nuclear fallout, a radiation suit (especially one with respiratory protection) may help reduce exposure to radioactive particles. This is especially important in the first hours or days after a nuclear event, when fallout levels are at their highest.

When Radiation Suits Are Not Necessary

Low-Level Radiation Areas If you are in an area with low to moderate radiation levels, such as an area that has been exposed to fallout but where radiation levels have decreased over time, a radiation suit may not be necessary. In these situations, staying indoors, using proper shielding, and limiting your exposure time are more effective strategies for minimizing radiation risks.

Sheltering Indoors For most people in a fallout zone, the best strategy is to remain indoors in a well-shielded building for the first 24 to 48 hours after a nuclear event. The primary concern in this scenario is not direct radiation exposure, but rather the presence of radioactive particles in the air. Sheltering in place with sealed windows and doors, along with HEPA filters or makeshift air filtration systems, can significantly reduce the need for a full-body radiation suit.

Long-Term Survival In the long term, once radiation levels have decreased and decontamination efforts are underway, it's more important to focus on minimizing exposure by avoiding contaminated areas, using proper hygiene (washing off any fallout particles from the skin), and limiting time spent outdoors. Simple protective measures, such as wearing masks, goggles, gloves, and clothing that covers exposed skin, can provide adequate protection in many cases.

Alternatives to Radiation Suits

If you don't have access to a professional radiation suit, there are still several measures you can take to protect yourself from radiation exposure in a fallout zone. These alternatives focus on minimizing contact with radioactive particles and reducing your exposure to harmful radiation:

Layered Clothing

Wearing multiple layers of clothing can provide some protection against radioactive dust and particles. While this won't block gamma radiation, it can help prevent alpha and beta particles from reaching your skin. Be sure to cover all exposed skin, including wearing gloves, long sleeves, pants, and closed shoes.

Clothing material: Synthetic materials like nylon or polyester can offer better protection than cotton, as they are less likely to allow particles to pass through.

Decontamination: After being outdoors in a fallout zone, remove and dispose of contaminated outer layers of clothing and wash thoroughly to remove any remaining radioactive particles from your skin.

Respirators or Masks

Respirators or high-quality masks are crucial for preventing the inhalation of radioactive dust. N95 or P100 respirators can filter out many of the particles in the air, while simple cloth masks provide some protection in lower-risk environments.

Goggles: In addition to a respirator, wearing goggles or safety glasses will protect your eyes from airborne radioactive particles.

Improvised Radiation Protection

If you need to make an improvised radiation suit, materials like plastic sheeting, trash bags, or ponchos can be used to create a temporary barrier against radioactive particles. Tape or seal the openings at the wrists, ankles, and neck to prevent dust from entering.

Plastic ponchos: A disposable plastic poncho, combined with gloves and a mask, can act as a makeshift radiation barrier when traveling through a fallout zone. After exposure, remove and discard the poncho and any other outer protective layers to avoid bringing contamination inside.

Other Protective Measures for Radiation

In addition to wearing radiation suits or improvised protective clothing, there are several key strategies for reducing your overall radiation exposure in a fallout scenario:

Time, Distance, and Shielding

The three key principles of radiation protection are time, distance, and shielding:

Time: Minimize the time you spend in radiation-affected areas. The less time you are exposed, the lower your overall radiation dose.

Distance: Increase your distance from the radiation source whenever possible. Radiation intensity decreases significantly with distance, so staying far away from contaminated zones can reduce your exposure.

Shielding: Place as much shielding (thick walls, earth, lead, or concrete) between yourself and the radiation source as possible. When sheltering indoors, seek out basements or interior rooms away from windows and doors.

Decontamination

After exposure to radioactive particles, it's critical to decontaminate as soon as possible:

Remove clothing: Carefully remove and discard contaminated clothing to eliminate up to 90% of radioactive particles from your body.

Wash skin: Use soap and water to thoroughly wash exposed skin and hair, paying special attention to areas like the face, hands, and feet.

Rinse eyes and nose: Use clean water or saline to rinse out your eyes and nose to remove any inhaled or airborne particles.

Potassium Iodide (KI) Tablets

Potassium iodide (KI) tablets can help protect your thyroid from radioactive iodine, which is a common byproduct of nuclear explosions. By saturating your thyroid with stable iodine, KI tablets prevent the gland from absorbing radioactive iodine, reducing the risk of thyroid cancer.

Dosage: Follow the recommended dosage instructions carefully, as too much iodine can be harmful. KI tablets should be taken shortly before or after exposure to radioactive iodine for maximum effectiveness.

Radiation suits can be valuable in specific high-risk situations, such as decontamination work or traveling through fallout zones, but they are not always necessary for survival in a post-nuclear world. In many cases, other protective measures—such as sheltering in place, wearing layered clothing, using masks and goggles, and minimizing exposure—can offer adequate protection against radioactive particles. Understanding the different types of radiation and knowing when and how to protect yourself is crucial for long-term survival. Whether you have access to professional radiation suits or need to rely on improvised protective gear, staying informed and prepared will help you minimize the dangers of radiation in a nuclear fallout scenario.

Foraging in Fallout Zones: What's Safe to Eat

In a post-nuclear world, food sources will become scarce, and foraging may be a critical survival skill. However, in fallout zones, the dangers of radiation contamination make foraging a highly risky endeavor. Radioactive particles can settle on plants, water, and soil, potentially contaminating food sources. Understanding what's safe to eat, how to identify contamination, and how to minimize risks is essential for survival. This chapter will cover strategies for foraging in fallout zones, what types of plants and food may be safer, and how to reduce the risks of consuming contaminated food.

The Risks of Foraging in Fallout Zones

In the aftermath of a nuclear disaster, radioactive fallout—composed of dust and particles—will settle on everything exposed to the environment, including plants, water, and animals. Radioactive contamination can affect foragers in two main ways:

Surface contamination: Radioactive dust and particles can settle on the leaves, fruits, and roots of plants, as well as on animals and water sources. Consuming contaminated food or water can lead to internal radiation exposure, which is extremely dangerous.

Absorption of radioactive isotopes: Plants and animals can absorb radioactive materials, such as cesium-137 or strontium-90, from contaminated soil and water. This internal contamination is more difficult to detect and cannot be washed away. Radioactive isotopes that mimic essential nutrients—such as cesium, which behaves like potassium, or strontium, which acts like calcium—can accumulate in plants and animals, making them hazardous to consume.

General Foraging Guidelines in Fallout Zones

Before discussing specific foods and plants, here are some general guidelines to follow when foraging in fallout zones. These guidelines are meant to minimize your exposure to radiation and help you make safer choices about what to eat.

Avoid Immediate Foraging After a Nuclear Event

For the first 24 to 48 hours after a nuclear explosion, radiation levels will be at their highest due to the fresh fallout. During this time, it is unsafe to forage or venture outdoors. Stay indoors or in a fallout shelter until radiation levels decrease significantly, which may take days or weeks, depending on the severity of the fallout.

Choose Sheltered Plants

Plants that have been sheltered from direct fallout—such as those growing under thick tree canopies, in caves, or inside structures—are less likely to be contaminated than those exposed directly to the air. Foraging in areas with natural cover can reduce the risk of surface contamination.

Underground plants: Root vegetables like carrots, potatoes, and onions may be safer because they grow underground, where they are less likely to be affected by surface fallout. However, they may still absorb radioactive elements from contaminated soil, so caution is necessary.

Harvest Edible Parts That Can Be Peeled or Shelled

Fruits, nuts, and vegetables with thick skins, shells, or husks may offer some protection against surface contamination. By peeling or removing the outer layer, you may reduce your exposure to radioactive dust. However, this does not guarantee that the food is safe, as internal contamination is still possible.

Examples: Fruits with thick peels, such as bananas, oranges, or melons, as well as nuts with hard shells, like walnuts or almonds, may be safer options for foraging.

Wash and Decontaminate Food Thoroughly

If you must consume foraged food, washing it thoroughly with clean, uncontaminated water can help remove radioactive particles from the surface. If you don't have access to clean water, you can use filtered or boiled water, though it may not be 100% effective at removing radioactive contaminants.

Washing technique: Soak the food in clean water and gently scrub with a brush to remove surface dust. Discard the washing water in a safe location, away from living areas and water sources.

Avoid Foraging Near Known Fallout Hotspots

Certain areas are more likely to have higher levels of fallout, including locations near the blast zone, downwind of the explosion, or areas where radioactive materials may have settled due to weather patterns. Avoid foraging near these hotspots, as plants and animals in these areas are likely to be heavily contaminated.

Hotspot examples: Cities, factories, military installations, or areas with nuclear reactors are more likely to be contaminated.

Identifying Safer Foraging Options

While no food in a fallout zone can be guaranteed to be entirely safe, certain types of plants and food sources may be less contaminated than others. Here are some specific categories of plants and animals to consider when foraging.

Root Vegetables

Root vegetables, which grow underground, may be safer than plants exposed directly to fallout, though they are not completely free from risk. These vegetables can absorb radioactive materials from the soil, particularly isotopes like cesium-137 and strontium-90, but the risk may be lower than with above-ground plants.

Safer options: Carrots, potatoes, turnips, beets, and onions.

Risk factors: If the soil is contaminated, root vegetables can still absorb radioactive isotopes. Peeling the skin and cooking them thoroughly may help reduce surface contamination, but it won't eliminate internal radiation.

Nuts and Seeds

Nuts and seeds with thick shells or husks provide some protection against surface contamination. The hard outer layer may shield the edible parts from direct fallout, making them safer to consume after peeling or shelling.

Safer options: Walnuts, almonds, sunflower seeds, and pumpkin seeds.

Risk factors: While the outer shell may protect the edible part, nuts and seeds can still absorb radioactive materials from contaminated soil or water.

Wild Edible Plants

Certain wild plants may be less affected by fallout if they are sheltered by trees or grow in areas that are not directly exposed to radioactive particles. However, the risk of contamination is still present, especially if the plants have absorbed radioactive elements from the soil.

Safer options: Sheltered plants, such as those growing under thick canopies or near water sources that are not contaminated, may be a better choice.

Risk factors: Avoid plants growing in open fields or areas with visible fallout or dust.

Berries and Fruits with Thick Peels

Berries and fruits can be contaminated by fallout, especially those growing on bushes or trees. However, fruits with thick peels that can be removed may offer some protection from surface contamination.

Safer options: Melons, bananas, oranges, avocados, and coconuts. Berries that can be washed thoroughly may also be an option, though they are more susceptible to surface contamination.

Risk factors: Thin-skinned fruits and berries may not provide enough protection from radioactive particles, making them more hazardous to eat without extensive washing.

Animals and Fish

While foraging for animals or fish may seem like a viable option for protein, the risks of consuming contaminated animals are high. Animals that graze on contaminated plants or drink from radioactive water sources can accumulate radiation in their bodies, particularly in their bones, organs, and muscles.

Safer options: Fish from uncontaminated freshwater sources may be safer than animals that graze on land. However, any animals or fish in areas exposed to fallout should be considered risky.

Risk factors: Larger animals that have consumed a significant amount of contaminated food or water will likely have high levels of internal radiation. Fish from contaminated rivers or lakes may also absorb radioactive isotopes.

Techniques for Reducing Radiation in Foraged Food

If you need to forage for food in a fallout zone, there are several steps you can take to reduce the risk of consuming contaminated items. While these methods cannot guarantee safety, they can help lower the levels of radioactive particles in food.

Peeling and Scraping

For fruits, vegetables, and root crops, removing the outer layers can reduce surface contamination. Peel fruits, scrape off the outer layer of root vegetables, and remove the outer leaves of plants like cabbage or lettuce. Discard the peelings in a safe location, away from water sources and living areas.

Cooking

Cooking food, particularly boiling, can help remove some radioactive contaminants from the surface. Boiling foraged food in clean water and discarding the water afterward may reduce the risk of surface contamination, though it won't eliminate internal radiation from plants or animals that have absorbed radioactive isotopes.

Boiling technique: Boil the food for several minutes in uncontaminated water, then discard the boiling water and rinse the food again in clean water before consuming.

Avoid Contaminated Water Sources

Water sources are highly vulnerable to fallout contamination, especially surface water like rivers, lakes, and ponds. When foraging, avoid collecting plants or animals near contaminated water sources. If you need to collect water, use a water filtration system specifically designed to remove radioactive particles.

Safe water techniques: Filter water using a high-quality water filtration system or boil water for several minutes to remove some contaminants. However, boiling alone may not be sufficient to remove all radioactive isotopes.

Test for Radiation

If you have access to a Geiger counter or other radiation detection equipment, you can test food, water, and soil for radiation levels before consuming or harvesting. This is the most effective way to assess the level of contamination, though it may not always be practical or available in survival situations.

Testing technique: Use a Geiger counter to check for high radiation readings on the surface of food or plants. If radiation levels are dangerously high, avoid consuming the food.

Foraging in fallout zones carries significant risks, as plants, animals, and water can be contaminated with radioactive particles. However, in a survival situation, foraging may be necessary. By following key guidelines—such as choosing sheltered plants, peeling and washing food thoroughly, and avoiding high-risk areas—you can reduce the risk of radiation exposure. While no food in a fallout zone can be considered completely safe, careful selection and preparation can help increase your chances of survival while minimizing exposure to harmful radiation. Always prioritize safety, and when possible, supplement foraging with stored food and other reliable sources of nourishment.

Fishing and Hunting in a Contaminated World

In a post-nuclear world, traditional methods of acquiring food, such as fishing and hunting, will become essential for survival, especially as agricultural systems collapse and access to stored food dwindles. However, radioactive contamination in the environment creates serious risks when sourcing food from wild animals and fish. Radiation can permeate ecosystems, affecting water sources, soil, plants, and, subsequently, the animals that rely on them for survival. This chapter will explore the dangers and considerations of fishing and hunting in a contaminated world, offering strategies to minimize exposure to radiation and ensure that the food you catch is as safe as possible to eat.

Understanding Radiation Contamination in Wildlife

Radiation affects the environment in several ways, and animals are not immune to its impact. Radioactive particles from fallout can settle on plants, water, and soil, becoming incorporated into the food chain. Animals that eat contaminated plants or drink from irradiated water sources can accumulate radiation in their tissues, organs, and bones. This process is known as bioaccumulation—where radioactive substances build up in an organism over time, making them more dangerous to consume.

There are two primary ways animals and fish can become contaminated by radiation:

External contamination: Radioactive dust and particles settle on the skin, fur, or feathers of animals and fish. This can often be washed away or cleaned before consumption, but it still poses a risk of exposure.

Internal contamination: The more dangerous form of contamination occurs when animals ingest radioactive particles through their food and water, which are then absorbed into their bodies. Radioactive isotopes like cesium-137 (which mimics potassium) and strontium-90 (which mimics calcium) can accumulate in the muscles, bones, and organs of animals. Internal contamination is difficult to detect without specialized equipment and cannot be removed through washing or cooking.

Risks of Eating Contaminated Animals and Fish

Consuming contaminated animals or fish can lead to internal radiation exposure, which occurs when radioactive particles are ingested and absorbed into your body. This type of exposure is particularly dangerous because the radiation can damage tissues and cells over time, increasing the risk of radiation sickness, cancer, and other long-term health problems.

Fish: Fish living in contaminated water can absorb radioactive isotopes through their gills and by consuming irradiated algae or smaller fish. Larger fish, which are higher up in the food chain, can accumulate more radiation due to bioaccumulation.

Game animals: Animals that graze on contaminated plants or drink from radioactive water sources will also accumulate radiation in their bodies, especially in their bones and organs. Larger animals may have higher levels of contamination because they consume more food and water.

Birds: Birds that feed on contaminated insects, seeds, or plants can also be affected, though their relatively short lifespans may limit the amount of radiation they accumulate compared to larger mammals.

Key Principles for Fishing and Hunting in Fallout Zones

Despite the risks, fishing and hunting may be necessary for survival in a post-nuclear world. By understanding where and how radiation affects wildlife, you can reduce the dangers associated with eating contaminated animals and fish. Below are key strategies for fishing and hunting in fallout zones.

Avoid Contaminated Water Sources

One of the most significant risks for fish contamination is the water they live in. Fallout can contaminate lakes, rivers, and ponds, especially those exposed to fallout clouds or runoff from contaminated soil. Avoid fishing in known fallout hotspots, particularly bodies of water located near blast zones or downwind from major explosions.

Safer options: Look for freshwater sources that are sheltered or underground, such as spring-fed streams, which may be less exposed to fallout. Flowing water, such as fast-moving rivers, may also be safer than stagnant bodies of water, as contamination can be diluted more quickly.

Testing water: If possible, test the water with a Geiger counter or other radiation detection equipment before fishing. High radiation levels in the water indicate that the fish are likely contaminated as well.

Select Smaller, Younger Animals and Fish

In both fishing and hunting, smaller and younger animals are generally less likely to be heavily contaminated than larger, older ones. This is because bioaccumulation occurs over time, so animals that have had longer exposure to contaminated food and water will carry more radioactive particles in their bodies.

Fish: Target smaller fish, which are lower in the food chain and have had less time to accumulate radiation. Avoid predatory fish like pike, bass, or large catfish, which consume other contaminated fish and accumulate higher levels of radiation as a result.

Game animals: Hunt smaller animals like rabbits, squirrels, or young deer, as they are less likely to have accumulated dangerous levels of radiation compared to larger, older animals like adult elk or moose.

Focus on Lean Meat

Radioactive isotopes like cesium-137 tend to accumulate in muscle tissue, while strontium-90 often concentrates in bones. When preparing game animals or fish, focus on eating lean muscle meat and avoid consuming organs, bones, or marrow, where higher concentrations of radiation may be found.

Avoid organ meats: Organs such as the liver, kidneys, and brain tend to accumulate higher levels of radioactive materials. In survival situations, it's best to avoid these parts and stick to lean muscle tissue.

Trim fat: Cesium-137 can also be stored in animal fat, so trim excess fat from the meat before cooking.

Wash and Clean Animals Thoroughly

Before consuming any game or fish, wash and clean them thoroughly to remove surface contamination. This is especially important for animals that may have been exposed to fallout on their fur, feathers, or scales.

Fish: Scale and gut fish before cooking. Wash them thoroughly in clean, uncontaminated water, and discard the internal organs where radiation may be concentrated.

Game animals: Skin and dress animals carefully, being mindful to avoid contact with the internal organs. Wash the meat in clean water and cook it thoroughly to kill any bacteria or parasites, though this won't remove internal radiation.

Cook Food Properly

Cooking meat and fish does not remove radioactive isotopes, but it can help reduce the risk of bacterial or parasitic infections, which are also a concern in survival scenarios. Boiling, roasting, or grilling are the best methods for preparing meat and fish in a fallout zone.

Boiling: Boiling fish or meat in clean water can help reduce surface contamination, though it won't eliminate internal radiation. Discard the boiling water afterward, as it may contain radioactive particles.

Avoid stews or broths: Stewing or making broth from bones, organs, or fatty cuts of meat can concentrate radiation, especially if the water used for cooking becomes contaminated.

Avoid Scavenging or Hunting Sick Animals

Radiation exposure weakens the immune systems of animals, making them more susceptible to illness. Avoid scavenging dead animals or hunting animals that appear sick or lethargic, as they may have accumulated dangerously high levels of radiation or be carrying secondary infections due to their weakened condition.

Signs of radiation sickness in animals: Hair loss, open sores, or unusual behavior may indicate that an animal is suffering from radiation sickness or other diseases. Avoid consuming these animals, as their meat may be highly contaminated.

Specific Considerations for Fishing in Contaminated Waters

Fishing in contaminated water presents unique challenges because fish can absorb radioactive materials directly through their gills as well as through their food. Here are some additional considerations when fishing in fallout zones:

Freshwater fish vs. saltwater fish: In a fallout scenario, freshwater fish may be more accessible, but they are also more vulnerable to contamination from local fallout. Saltwater fish, particularly those from deep ocean areas, may be less affected by localized fallout, but accessing them may be difficult in a survival scenario.

Fish species: Smaller, fast-growing fish like trout, perch, and bluegill may be safer to eat than larger, slower-growing species. Predatory fish that eat other contaminated fish, such as bass or pike, should be avoided.

Testing fish: If you have access to a Geiger counter, test the fish for radiation before consuming. Fish with higher-than-normal readings should be avoided, especially if they are from slow-moving or stagnant water.

Specific Considerations for Hunting in Fallout Zones

When hunting game animals in contaminated environments, the risks of radiation exposure vary depending on the species, their diet, and their habitat. Animals that graze on contaminated plants or drink from irradiated water sources are at higher risk of contamination. Here are some additional tips for hunting in fallout zones:

Grazing vs. browsing animals: Grazing animals like deer, elk, and cattle that feed on grass or ground-level vegetation are more likely to consume fallout-contaminated plants. Browsing animals, such as goats or moose, that feed on leaves and branches from higher up may have lower levels of contamination.

Insectivores and omnivores: Animals that eat insects or a varied diet may be less contaminated than strict herbivores or carnivores. However, insectivores like birds that consume radioactive insects should still be approached with caution.

Water sources: Avoid hunting animals that rely on contaminated water sources, such as lakes, rivers, or ponds, especially if they are located in fallout-heavy areas.

Safety and Decontamination After Hunting or Fishing

After catching fish or game animals, it's essential to take steps to protect yourself from exposure to radiation and contamination:

Protective gear: Wear gloves, masks, and protective clothing when cleaning or handling game animals or fish to avoid contact with radioactive materials. After handling, wash your hands, tools, and clothing thoroughly.

Waste disposal: Discard contaminated parts of the animal, such as organs or bones, in a safe location away from water sources, living areas, or gardens.

Decontaminate yourself: After hunting or fishing, clean yourself and your gear thoroughly. Radiation particles can settle on your skin, hair, and clothing, so it's important to wash up with clean, uncontaminated water.

Fishing and hunting in a contaminated world come with significant risks, but they may be necessary for survival in a post-nuclear environment. By following best practices—such as avoiding fallout hotspots, selecting smaller or younger animals, cleaning and cooking food thoroughly, and focusing on lean meat—you can minimize your exposure to radiation. While no method is foolproof, these strategies will help you make safer choices when sourcing food in a world affected by nuclear fallout. Ultimately, the key to survival is caution, preparedness, and the ability to adapt to a dangerous and uncertain environment.

Adapting to a New Climate: Surviving the Cold of Nuclear Winter

In the aftermath of a large-scale nuclear conflict, the world could be plunged into a "nuclear winter," a period of drastically reduced temperatures caused by the atmospheric fallout from nuclear explosions. Massive amounts of soot, ash, and smoke from firestorms would rise into the upper atmosphere, blocking sunlight and reducing the earth's temperature by several degrees. The result would be a significant and prolonged drop in temperature, akin to a harsh and extended winter, even in typically warmer climates. Adapting to these new, colder conditions will be critical for survival, especially when food production, energy supplies, and shelter are already strained. This chapter will explore the impacts of nuclear winter on climate, how to adapt to the cold, and practical strategies for maintaining warmth and health in a world gripped by nuclear winter.

What is Nuclear Winter?

Nuclear winter is a theoretical scenario that describes the drastic cooling of the Earth's climate following a large-scale nuclear war. The intense heat from nuclear detonations would ignite widespread firestorms in cities, forests, and industrial areas. These firestorms would release massive amounts of soot, ash, and smoke into the atmosphere, where they could block sunlight for months or even years. Without adequate sunlight, global temperatures would drop, crops would fail, and ecosystems would collapse, creating a harsh environment for human survival.

The Impact of Nuclear Winter on Climate

A nuclear winter could have several significant impacts on the global climate, creating a much harsher and more challenging environment for survivors:

Global cooling: The primary effect of nuclear winter is a dramatic reduction in temperatures. Studies suggest that average global temperatures could drop by several degrees, leading to colder-than-usual conditions across the planet. This cooling would be most severe in regions near the nuclear blasts but could affect the entire globe.

Extended winters: Even in areas where winter typically lasts for only a few months, nuclear winter could extend these colder conditions for years. Summer seasons may be nonexistent or drastically shortened, leading to continuous cold weather for extended periods.

Reduced sunlight: The soot and ash in the upper atmosphere would block sunlight, reducing the amount of solar energy reaching the Earth's surface. This reduction in sunlight would not only cool the planet but also disrupt food production by limiting photosynthesis.

Severe weather patterns: With the climate destabilized, weather patterns could become more unpredictable. Blizzards, storms, and extreme cold could become more frequent, and some areas might experience harsher winters than others.

Adapting to the Cold: Survival Strategies

In a nuclear winter, survival will depend on your ability to adapt to the cold and maintain warmth in a world where energy and resources are limited. Below are key strategies for surviving the extreme cold of nuclear winter.

Building and Maintaining Warm Shelters

The first line of defense against the cold is ensuring you have adequate shelter. Whether you are in a home, bunker, or improvised shelter, insulating your living space and maintaining warmth is critical.

Insulation: Add extra insulation to your shelter to retain heat. If you're in a house, cover windows with blankets, plastic sheeting, or heavy curtains to keep drafts out. Line walls and doors with blankets, cardboard, or foam padding to trap heat inside.

Underground shelters: If you have access to an underground bunker or basement, it can be more effective at retaining heat due to the natural insulation of the earth. Underground shelters are also less exposed to the external environment and can protect you from fallout.

Seal off unused rooms: If you're sheltering in a house, limit the number of rooms you use to conserve heat. Seal off unused rooms with blankets, plastic, or furniture to trap heat in the areas where you spend most of your time.

Ventilation and air quality: While sealing off your shelter for warmth, remember to maintain adequate ventilation. In enclosed spaces, the buildup of carbon monoxide or carbon dioxide from heating sources can be dangerous. Use a carbon monoxide detector and ensure there is some airflow, even in tightly insulated shelters.

Heating Sources

Traditional heating sources may not be available during a nuclear winter due to fuel shortages, power outages, or damage to infrastructure. Finding alternative ways to generate heat is crucial for survival.

Wood-burning stoves: If you have access to a wood-burning stove, it can be one of the most reliable and long-lasting sources of heat during a nuclear winter. Ensure your chimney is clear of debris, and stockpile firewood in a dry, protected area. Wood stoves can also be used for cooking and boiling water, making them versatile survival tools.

Improvised heating: In the absence of a stove or furnace, improvised heating solutions like small fires, candle heaters, or even body heat can be used. For instance, a terracotta pot heater, created by placing a clay pot over a few candles, can radiate heat in a small space.

Solar heating: Although sunlight may be reduced, it may still be possible to use passive solar heating during the day. Use large windows or south-facing walls to capture any available sunlight, and use reflective surfaces (like aluminum foil) to direct heat into your living space.

Fuel alternatives: If you have access to fuel sources like propane, kerosene, or butane, these can be used to power small heaters or stoves. However, ensure that any fuel-burning devices are well-ventilated to prevent dangerous fume buildup.

Layering Clothing and Bedding

Proper clothing and bedding are essential for retaining body heat in cold conditions. In a nuclear winter, layering is one of the most effective ways to stay warm.

Layer your clothing: Wear multiple layers of clothing, starting with moisture-wicking materials like wool or synthetic fibers to keep sweat away from your skin. Add insulating layers such as fleece, down, or wool, and finish with a windproof and waterproof outer layer to protect against drafts and moisture.

Keep extremities warm: Pay extra attention to keeping your hands, feet, and head warm, as these areas are most vulnerable to heat loss. Wear insulated gloves, wool socks, and a thermal hat. Balaclavas or scarves can also help trap heat around your neck and face.

Bedding: At night, use multiple layers of blankets or sleeping bags to trap heat. Wool or down blankets are particularly effective at providing insulation. If you're sleeping in a cold room, wrap yourself in thermal blankets and use hot water bottles to provide additional warmth.

Food and Hydration in Cold Conditions

Cold weather increases the body's need for calories to maintain heat. Eating nutrient-dense, high-calorie foods and staying hydrated are critical for surviving the cold.

High-calorie foods: Focus on consuming foods high in fat, protein, and carbohydrates. Nuts, dried fruits, grains, canned meats, and legumes are excellent options for sustaining energy. In a nuclear winter, foraging or gardening may be difficult, so it's important to rely on stored food.

Warm meals: Whenever possible, eat warm meals to help maintain body heat. Soups, stews, and hot drinks can provide both nourishment and warmth. If you have access to a wood-burning stove or alternative heat source, cooking warm meals should be a priority.

Hydration: Cold weather can make it easy to forget to drink water, but dehydration is still a concern. Drink plenty of fluids, and if your water supply is limited, melt snow or ice for drinking water (but be sure to boil it to remove any contaminants). Avoid alcohol, as it can lower your body temperature and increase the risk of hypothermia.

Physical Activity to Maintain Warmth

Staying physically active can help generate body heat, especially if you're sheltering in a cold environment. Regular movement will improve circulation and help prevent frostbite or hypothermia.

Short bursts of activity: Perform short, moderate exercises like squats, push-ups, or stretching to increase your body temperature. Avoid excessive sweating, as this can cause you to lose heat more quickly when you cool down.

Group warmth: If you are sheltering with others, huddle together to conserve heat. Sharing body warmth, particularly at night, can make a significant difference in cold environments.

Preparing for the Long-Term Effects of Cold Weather

Nuclear winter may last for months or even years, depending on the severity of the conflict and the amount of soot and ash in the atmosphere. Long-term cold weather survival will require careful planning and resource management.

Stockpile firewood and fuel: Gather and store as much firewood, coal, or other heating fuel as possible before nuclear winter begins. Firewood should be kept dry and protected from the elements, as wet wood will burn inefficiently and produce less heat.

Learn cold-weather skills: If you live in a region that doesn't typically experience harsh winters, it's important to learn essential cold-weather survival skills, such as starting fires, building insulated shelters, and recognizing the signs of frostbite and hypothermia.

Mental health and morale: The psychological toll of living in a nuclear winter, combined with isolation and harsh conditions, can affect your mental health. Maintaining a routine, staying physically active, and having social interactions (even within your shelter) can help maintain morale. Simple activities like reading, storytelling, or playing games can keep your mind engaged during long, cold days.

Health Risks in Cold Environments

Cold weather brings unique health risks, particularly in a survival scenario where resources are limited. Understanding and mitigating these risks will be essential for staying alive in nuclear winter.

Hypothermia

Hypothermia occurs when your body loses heat faster than it can produce it, causing your core temperature to drop dangerously low. It can happen even in relatively mild cold if you're wet or exposed to the wind for extended periods.

Signs of hypothermia: Shivering, confusion, drowsiness, slurred speech, and shallow breathing. If untreated, hypothermia can be fatal.

Prevention: Stay dry, wear insulated clothing, and seek shelter from the wind. If hypothermia sets in, move the person to a warmer environment, remove wet clothing, and use blankets or body heat to warm them slowly.

Frostbite

Frostbite occurs when skin and tissue freeze due to extreme cold. It most often affects the fingers, toes, nose, and ears, and can lead to permanent tissue damage.

Signs of frostbite: Numbness, tingling, or pain in the affected area, followed by hard, pale, and cold skin. Severe frostbite can turn skin black and cause blisters.

Prevention: Keep extremities warm and covered, avoid prolonged exposure to the cold, and stay active to maintain circulation.

Respiratory Problems

Breathing cold air can irritate your respiratory system, especially if you're already exposed to fallout or ash particles. Prolonged exposure to cold air can lead to respiratory infections, bronchitis, or even pneumonia.

Prevention: Use a scarf or mask to warm the air before it enters your lungs. Avoid breathing in ash or fallout by staying indoors during high-risk periods and using air filters or masks if necessary.

Adapting to the cold of nuclear winter requires careful planning, resource management, and practical survival skills. Stay aware of the health risks associated with cold exposure, including hypothermia and frostbite, and take proactive steps to stay warm, dry, and nourished.

Building Community: The Importance of Group Survival

Surviving in a post-nuclear world is not a task easily accomplished alone. While self-reliance and individual survival skills are crucial, the reality is that humans are social beings, and building a community can significantly increase the chances of long-term survival. In the aftermath of a nuclear disaster, banding together with others not only increases the available pool of resources, skills, and knowledge, but also provides emotional support and protection from external threats. This chapter will explore the importance of group survival, how to build a cohesive community in a post-nuclear environment, and the practical steps for maintaining order, security, and cooperation in the face of extreme challenges.

Why Group Survival Is Key in a Post-Nuclear World

There are several critical reasons why surviving as part of a group is more advantageous than attempting to go it alone in a post-nuclear world. These include:

Shared resources: No individual can stockpile everything they need for long-term survival. Grouping with others allows for the pooling of food, water, medical supplies, tools, and equipment, ensuring that everyone has access to what they need. Communities can more effectively manage rationing and ensure that resources are allocated in a way that maximizes their usefulness.

Diverse skills: Survival requires a wide variety of skills—medical knowledge, mechanical skills, farming expertise, security tactics, and more. No single person can be an expert in every area. A group offers a broader range of skills, allowing members to take on specific roles that they are most proficient in. This specialization improves efficiency and the ability to tackle complex tasks.

Increased security: In a post-nuclear world, external threats such as hostile groups, looters, or wild animals may pose significant dangers. A community can more effectively organize defenses, guard rotations, and patrols to protect against these threats. Larger numbers mean greater strength in defending resources and people.

Psychological support: The mental strain of surviving a nuclear catastrophe can be overwhelming. Isolation can lead to anxiety, depression, and hopelessness. Being part of a community provides emotional support and helps maintain morale, which is vital for long-term survival. Social interaction, shared responsibility, and a sense of purpose within the group can stave off the psychological toll of a post-disaster environment.

Coordinated survival strategies: A group can coordinate survival efforts in a way that an individual cannot. This includes organized hunting or fishing trips, farming, foraging, maintaining infrastructure like water filtration systems, or repairing shelters. With a clear division of labor, the community can focus on multiple survival tasks simultaneously, increasing overall efficiency and chances of success.

Steps to Building a Post-Nuclear Survival Community

Building a community in a post-nuclear environment requires careful planning, trust, and cooperation. Whether you are gathering with family, neighbors, or fellow survivors, the key to successful community building lies in creating a functional group with shared goals and responsibilities. Below are practical steps for establishing a strong survival community.

Start with Trustworthy People

The foundation of any successful survival community is trust. In a post-nuclear world, where resources are scarce and tensions can run high, it's crucial to form a group with people you trust. These could be family members, close friends, neighbors, or individuals who share similar values and a commitment to survival.

Pre-existing relationships: If possible, start with people you already know and trust. Family and close friends are natural allies in a survival scenario, as they already have a vested interest in your well-being. Trust allows for smoother decision-making and cooperation, especially when conflicts arise.

Screen new members carefully: If you must invite strangers into your community, vet them carefully. Evaluate their skills, attitude, and willingness to contribute to the group. People who are reliable, cooperative, and skilled will be valuable assets, but be cautious of those who may disrupt group cohesion or act selfishly.

Define Roles and Responsibilities

Clear roles and responsibilities are essential for ensuring that the community functions smoothly. In a post-nuclear world, every member of the group must contribute to survival efforts. Defining specific roles helps prevent conflict and ensures that essential tasks are not overlooked.

Skills assessment: Begin by assessing the skills of each group member. Some may have medical training, others may have experience in farming or carpentry, and some may be skilled at hunting or security. Assign roles based on these skills to ensure everyone can contribute in meaningful ways.

Division of labor: Divide responsibilities into categories such as food production, security, medical care, water management, and shelter maintenance. Rotate tasks when appropriate to avoid burnout, especially for physically or emotionally taxing roles like security patrols.

Leadership: Leadership in a survival community should be based on expertise, trust, and consensus rather than brute force or authoritarianism. Identify individuals who are capable of leading specific areas, such as organizing food production, handling disputes, or planning defense strategies.

Establish Rules and Conflict Resolution Mechanisms

Every community, especially in a high-stress environment like a post-nuclear world, will encounter disagreements and conflicts. Establishing clear rules and conflict resolution methods from the outset can prevent small disputes from escalating into dangerous situations.

Basic rules: Create rules that cover essential survival behaviors, such as how resources will be shared, how disputes will be handled, and the consequences for hoarding or theft. These rules should be agreed upon by all members to ensure they are respected.

Conflict resolution: Set up a system for resolving conflicts. This could include group discussions, mediation by a neutral party, or a vote among community members. Having a clear process in place will prevent arguments from spiraling out of control, which could destabilize the group.

Accountability: Ensure that there are consequences for rule-breaking. This could range from loss of privileges within the community to expulsion for serious offenses, like theft or violence. However, balance punishment with fairness, as overly harsh consequences can lead to resentment and further conflict.

Develop a Resource Management Plan

Resources like food, water, fuel, and medical supplies will be limited in a post-nuclear world. Managing these resources effectively is critical for group survival. A well-organized plan ensures that the community can survive for the long term without running out of essential supplies.

Rationing: Establish a rationing system to ensure that food, water, and other essentials are distributed fairly. Rationing can be based on the number of people in the group and the availability of resources. Track consumption carefully and adjust rations as needed to stretch resources over time.

Stockpiling: Whenever possible, work together to increase the group's stockpile of essential supplies. This may involve organizing group foraging trips, hunting or fishing expeditions, or bartering with other communities for needed goods.

Renewable resources: Focus on sustainability by developing renewable resource systems. This could include setting up a communal garden, raising livestock, or building water collection and filtration systems. Renewable systems help ensure that the group isn't solely reliant on finite stockpiles.

Security and Defense

One of the primary advantages of group survival is the ability to defend against external threats. In a post-nuclear world, these threats could come from hostile groups, looters, or even desperate individuals looking to steal supplies. Establishing a security plan is vital for protecting the community.

Organized defense: Create a rotating schedule of security patrols to monitor the perimeter of your shelter or community area. Assign trusted individuals to guard duty, and ensure that all members are trained in basic self-defense and the use of weapons if necessary.

Physical defenses: Reinforce your shelter with physical barriers like fences, barricades, or makeshift walls. Use natural features like hills, rivers, or dense vegetation to your advantage when selecting or fortifying your community's location.

Communication and coordination: Establish a reliable method of communication within the community in case of an attack or external threat. This could involve hand-crank radios, whistles, or pre-arranged signals. Make sure everyone knows the plan and can respond quickly to potential dangers.

Foster a Sense of Community and Cooperation

Surviving in a post-nuclear world requires not just practical skills but also a strong sense of community and cooperation. Fostering group cohesion and maintaining positive morale is crucial for long-term survival.

Daily routines: Establish daily routines that give structure to the community's activities. Routines help maintain a sense of normalcy and prevent chaos. This could include set times for meals, work assignments, and community meetings.

Social activities: Encourage social interactions and community-building activities, such as shared meals, storytelling, or games. These activities can strengthen bonds between group members, reduce stress, and provide emotional relief in an otherwise bleak environment.

Mutual aid: Promote a culture of mutual aid, where members support one another in difficult tasks or during times of illness or injury. This fosters trust and helps ensure that everyone is cared for, which in turn strengthens the group's overall resilience.

Establish Communication with Other Groups

In the long term, it may be beneficial for your survival community to establish contact with other groups. Trade, cooperation, and sharing information about dangers or resources can increase the group's chances of survival.

Bartering and trade: If your community has surplus supplies, bartering with nearby groups for needed items can help diversify your resources. Establish safe zones or neutral areas where trading can occur without conflict.

Alliances: Forming alliances with nearby groups can provide mutual protection and support in times of need. Alliances can also create opportunities for shared labor, such as large-scale farming or construction projects that benefit both communities.

Radio communication: If available, use ham radios, walkie-talkies, or other communication devices to keep in contact with neighboring communities. Share information about radiation levels, safe zones, and available resources to improve everyone's chances of survival.

Maintaining Long-Term Stability in a Survival Community

As the months and years of post-nuclear survival progress, maintaining stability in your community will become increasingly challenging. Issues such as resource depletion, leadership disputes, and interpersonal conflicts can threaten the group's cohesion. Here are some tips for maintaining long-term stability:

Rotate leadership roles: To prevent power struggles and maintain fairness, consider rotating leadership roles periodically. This allows different group members to take on responsibilities and ensures that no one person becomes too dominant or controlling.

Encourage problem-solving: Foster a group culture where problem-solving is a shared responsibility. When challenges arise, encourage group discussions and brainstorming sessions to find solutions collectively. This can prevent feelings of isolation or frustration among group members.

Adapt to changing circumstances: Survival in a post-nuclear world will require constant adaptation. Be prepared to change plans, adjust resources, or modify community rules as new challenges arise. Flexibility and adaptability are key to long-term survival.

Prevent burnout: Physical and emotional exhaustion can lead to burnout, which in turn can cause poor decision-making and conflict. Encourage rest, social interaction, and emotional support to help group members avoid burnout.

Building and maintaining a survival community in a post-nuclear world is crucial for long-term survival. A well-organized group with defined roles, shared resources, and a strong sense of cooperation will fare much better than individuals attempting to survive alone. By pooling skills, managing resources effectively, and establishing clear rules for security and cooperation, your community can become a safe haven in an otherwise dangerous and chaotic environment. Strong social bonds and mutual support will be the foundation of group survival, providing both physical and psychological resilience in the face of adversity.

Staying Informed: Understanding Radiation News and Updates

In a post-nuclear world, staying informed about radiation levels and potential dangers in your environment will be crucial for survival. Understanding radiation news and updates—whether through official broadcasts, local reports, or scientific readings—is key to making decisions about where and when to travel, which areas are safe, and how to avoid exposure to harmful levels of radiation. This chapter will explore how to stay informed about radiation levels, interpret radiation news, and gather updates from reliable sources. It will also cover how to use radiation detection tools to supplement external reports, ensuring that you and your community can respond quickly and appropriately to changing conditions.

The Importance of Staying Informed About Radiation

After a nuclear disaster, the environment will be contaminated with radioactive fallout, which can linger in the atmosphere, soil, water, and food supplies for years. Radiation levels can fluctuate over time and vary depending on your location. Being aware of current radiation levels and understanding updates about fallout zones can help you make critical decisions for your safety. Whether you're sheltering in place, planning a journey, or managing a community, you'll need accurate, up-to-date information to minimize radiation exposure.

Key reasons why staying informed about radiation is essential:

Avoiding high-risk areas: Some regions will be more heavily contaminated than others, especially areas near blast zones or where fallout clouds have settled. Knowing which areas have high radiation levels helps you avoid unnecessary exposure.

Monitoring radiation decay: Over time, radiation levels will decrease as radioactive isotopes decay. Staying informed about these changes can help you determine when it's safe to move through previously hazardous areas.

Planning travel and evacuation: If you need to move from one location to another, knowing the current radiation levels along your route is vital. Travel through areas with high radiation levels could result in dangerous exposure.

Protecting food and water supplies: Radiation can contaminate crops, water sources, and livestock. Updates on radiation levels in specific regions can inform your choices about foraging, farming, and gathering water.

Sources of Radiation News and Updates

In the immediate aftermath of a nuclear event, traditional news sources may be disrupted, but there are still ways to stay informed. Radio broadcasts, military or government updates, and local monitoring systems may provide essential information about radiation levels and fallout patterns.

Radio Broadcasts

One of the most reliable ways to receive radiation updates after a nuclear disaster is through emergency radio broadcasts. Many governments and emergency services are equipped to transmit information about radiation levels, weather patterns, and safe zones via AM, FM, or shortwave radio frequencies. Hand-crank or battery-powered radios are essential survival tools for accessing these broadcasts.

Emergency Alert System (EAS): In some countries, the government may activate the Emergency Alert System (or its equivalent) to provide regular updates on radiation, fallout zones, and safe areas. These updates are typically broadcast on local AM or FM radio stations.

Shortwave radio: If local radio stations are down, shortwave radio broadcasts may provide information from more distant regions or international sources. These broadcasts can be received over vast distances, making them valuable in situations where local infrastructure has collapsed.

Ham radio: Amateur or ham radio operators often serve as a crucial information network in disaster situations. If you or someone in your community is a licensed ham radio operator, you can access valuable information about radiation levels, communicate with other survivors, and receive updates from government or military channels.

Government and Military Updates

In many countries, government and military agencies will be responsible for monitoring radiation levels and disseminating updates to the public. These agencies often have the most advanced radiation detection equipment and can provide accurate, up-to-date information on which areas are safe or dangerous.

Civil defense networks: In the event of a nuclear disaster, civil defense organizations may be activated to provide information to survivors. These networks may use radio, public address systems, or even community bulletin boards to keep people informed about radiation levels and safe zones.

Military monitoring: Military units tasked with disaster response often have access to advanced radiation detection equipment. If the military is present in your area, they may provide radiation monitoring and updates on local conditions. In some cases, they may also establish safe zones or evacuation routes based on current radiation readings.

Local Monitoring Networks

Local communities may set up their own radiation monitoring systems using portable Geiger counters or dosimeters. These networks allow individuals or small groups to track radiation levels in their immediate area and share updates with others.

Community-based monitoring: In survival situations, communities may pool resources to create local monitoring stations. Volunteers can use portable radiation detectors to take readings and report back to the group on radiation levels in different areas. This information can be shared via word of mouth, written reports, or small-scale radio systems.

Mobile monitoring teams: Some communities may organize mobile teams equipped with radiation detection devices to monitor key locations such as water sources, food storage areas, or potential travel routes. These teams can provide regular updates on local radiation conditions.

Using Radiation Detection Tools

While external sources of radiation news and updates are valuable, it's essential to have your own tools for detecting and measuring radiation in your immediate environment. This allows you to verify the accuracy of the information you receive and make decisions based on real-time conditions. Two of the most common tools for detecting radiation are Geiger counters and dosimeters.

Geiger Counters

A Geiger counter is a handheld device that detects radiation levels in the environment. It works by measuring ionizing radiation, such as alpha, beta, and gamma rays. Geiger counters give real-time readings, usually in microsieverts per hour (μSv/h), which can help you determine the safety of your surroundings.

How to use a Geiger counter: Simply point the Geiger counter at the area you want to test—such as soil, water, food, or the air—and watch the reading on the display. The device will provide an immediate measurement of radiation levels. Geiger counters are especially useful for checking radiation levels before entering a new area or consuming foraged food.

Interpreting Geiger counter readings: Radiation levels are typically measured in microsieverts per hour. A reading below 0.1 μSv/h is generally considered safe. Levels between 0.1 and 1 μSv/h may pose a long-term risk if exposure continues for days or weeks. Levels above 1 μSv/h indicate dangerous conditions where exposure should be minimized or avoided.

Dosimeters

While Geiger counters measure radiation in the environment, dosimeters measure the cumulative amount of radiation a person has been exposed to over time. This is important because the risk of radiation poisoning increases with cumulative exposure, even at lower levels of radiation.

How to use a dosimeter: Wear the dosimeter on your body throughout the day. The device will track the total amount of radiation exposure you've received. At the end of each day, check the reading to see how much radiation you've accumulated.

Interpreting dosimeter readings: Dosimeters measure total exposure in millisieverts (mSv). A cumulative dose of less than 100 mSv over a year is considered low risk. Doses between 100 and 250 mSv increase the risk of cancer and other long-term health effects, while doses above 250 mSv can cause radiation sickness if absorbed over a short period.

Supplementing with Air Filters and Masks

In addition to monitoring radiation levels with Geiger counters and dosimeters, it's important to reduce your exposure to radioactive particles in the air. Fallout can contain dangerous particles that can be inhaled, leading to internal contamination.

Air filters: If possible, use air filters with HEPA (High-Efficiency Particulate Air) filters to remove radioactive particles from the air in your shelter. These filters can capture particles down to a very small size, reducing your exposure to airborne radiation.

Masks: Wear N95 or P100 respirators when traveling through areas with high fallout. These masks are designed to filter out fine particles, including radioactive dust. They won't protect you from gamma radiation, but they can reduce the risk of inhaling radioactive particles.

Interpreting Radiation Updates and Making Decisions

Once you have access to radiation news, updates, and data from your detection tools, the next step is interpreting that information and making decisions to protect yourself and your community. Here's how to make sense of radiation updates and use that information for survival.

Understanding Safe vs. Dangerous Radiation Levels

Radiation levels will vary depending on your location, the distance from the blast zone, and the time since the nuclear event. Understanding what constitutes a safe or dangerous level of radiation is key to minimizing exposure.

Safe levels: In general, radiation levels below 0.1 μSv/h are considered safe for long-term exposure. Levels between 0.1 and 1 μSv/h are manageable for short periods but may pose health risks if exposure is prolonged over days or weeks.

Dangerous levels: Radiation levels above 1 μSv/h are considered hazardous, especially if sustained over several hours. Prolonged exposure to levels above 1 μSv/h increases the risk of radiation poisoning and long-term health effects. In these cases, limit your time outdoors or in exposed areas and seek shelter immediately.

Acute radiation doses: In cases where radiation levels spike (above 10 μSv/h or higher), avoid exposure as much as possible. Short-term exposure to high levels of radiation can lead to acute radiation sickness, which can be life-threatening.

Using Updates for Travel Planning

When planning travel or evacuation routes, radiation updates are crucial for determining which areas are safe to pass through and which should be avoided. Use the following strategies to plan safe travel:

Avoid hot zones: If radiation updates indicate high levels of fallout in specific areas (often referred to as "hot zones"), avoid traveling through these regions. Even short exposure in a hot zone can be dangerous.

Monitor decay: Radiation levels will gradually decrease over time due to the decay of radioactive isotopes. Updates that show declining radiation levels can help you determine when it's safe to enter previously contaminated areas.

Check along the route: If you're planning a journey, gather radiation updates for the entire route, not just the starting point and destination. Radiation levels can vary significantly between locations, so it's important to know the conditions along the way.

Responding to Radiation Spikes

In the days, weeks, and months following a nuclear event, it's possible that radiation levels may spike due to weather patterns, new fallout, or other factors. Staying informed about these spikes allows you to take action to protect yourself.

Sheltering in place: If a radiation spike is reported in your area, shelter in place until levels decrease. Close all windows and doors, seal vents with plastic or duct tape, and use air filters if possible. Avoid going outside during radiation spikes unless absolutely necessary.

Wait for clear reports: If you're unsure about radiation levels, wait for official reports or use your own detection tools to confirm whether the spike is temporary or ongoing. Don't rush back into high-exposure areas without verifying that it's safe.

Staying informed about radiation levels and updates is critical for survival in a post-nuclear world. By using radio broadcasts, government or military updates, and community monitoring systems, you can stay aware of changing conditions and make informed decisions about your safety. Supplementing this information with your own radiation detection tools—such as Geiger counters and dosimeters—provides real-time data that helps protect you from dangerous exposure. Whether you're planning travel, sheltering in place, or monitoring food and water supplies, understanding and interpreting radiation news and updates will give you the knowledge you need to navigate a hazardous environment and ensure your survival in the long term.

The Role of Government in Post-Nuclear Survival

In the aftermath of a nuclear disaster, the role of government in providing guidance, resources, and order can be a critical factor in whether society survives and rebuilds or descends into chaos. Governments are responsible for organizing emergency responses, managing resources, coordinating relief efforts, and ensuring the safety and well-being of survivors. However, in a post-nuclear world, the infrastructure, leadership, and capacity of governments may be severely compromised, leaving individuals and communities to question how much they can rely on governmental aid. This chapter explores the role of government in post-nuclear survival, the limitations and challenges governments may face, and how survivors can navigate government involvement while remaining self-sufficient.

The Immediate Role of Government After a Nuclear Event

In the immediate aftermath of a nuclear event, the government's role is to protect citizens, manage the fallout, and stabilize the situation as quickly as possible. This typically involves activating emergency response systems, organizing evacuations, and disseminating critical information about radiation levels, safe zones, and fallout risks. The ability of the government to respond effectively depends on how intact its infrastructure remains and the scale of the nuclear disaster.

Emergency Response and Evacuation

One of the primary roles of government after a nuclear event is to organize and implement emergency evacuations for people living in or near the blast zone or fallout areas. This involves identifying the safest evacuation routes, establishing temporary shelters, and providing transportation for those who cannot evacuate on their own.

Evacuation zones: Governments may designate specific evacuation zones based on the proximity to the blast and expected fallout patterns. Citizens will be directed to move away from high-risk areas and head toward pre-established shelters or safe zones, often located far from the fallout path.

Shelters: In some cases, governments may have already built public fallout shelters in preparation for such disasters. These shelters are typically designed to protect people from radioactive fallout and provide basic necessities like food, water, and medical care.

Dissemination of Critical Information

The government's ability to communicate accurate and timely information is vital for helping citizens avoid dangerous radiation exposure and make informed decisions. This includes broadcasting radiation levels, advising people to shelter in place or evacuate, and providing updates on areas affected by fallout.

Emergency broadcasts: Governments will often use radio, television, and public address systems to broadcast emergency information. These broadcasts are designed to inform the public about the safest course of action, where to find shelter, and how to protect themselves from radiation.

Radiation monitoring: Many governments have radiation monitoring systems in place to track fallout and contamination levels. This data is essential for advising people on where it is safe to travel and when it is safe to return to their homes.

Law Enforcement and Order Maintenance

In the chaotic period following a nuclear disaster, maintaining law and order becomes a critical government responsibility. Looting, violence, and civil unrest can break out as resources become scarce and desperation sets in. Law enforcement, including local police, military forces, and national guards, may be mobilized to prevent criminal activity and protect survivors.

Curfews and restrictions: Governments may impose curfews or restrictions on movement to prevent looting, ensure public safety, and control the flow of people in and out of contaminated areas.

Martial law: In extreme cases, martial law may be declared, giving military forces the authority to take control of civilian functions to restore order and manage the disaster. Under martial law, civilians may be subject to strict regulations on movement, curfews, and the distribution of resources.

The Long-Term Role of Government in Post-Nuclear Survival

Once the immediate crisis is under control, governments must shift their focus to long-term survival and recovery. This involves organizing the rebuilding of infrastructure, providing food and medical supplies, coordinating relief efforts, and helping citizens navigate the long-term health risks of radiation exposure.

Resource Distribution and Management

The scarcity of essential resources like food, clean water, medical supplies, and fuel will be one of the most pressing challenges in a post-nuclear world. Governments will likely play a central role in managing and distributing these resources to ensure that the population has access to basic necessities.

Rationing systems: Governments may implement rationing systems to prevent hoarding and ensure that resources are distributed fairly. Citizens may be given ration cards or tokens to access food, water, and medical supplies at government distribution centers.

Supply chains: The government will need to re-establish supply chains to bring in necessary resources from unaffected areas or international partners. This includes coordinating shipments of food, medicine, and fuel to affected regions and working with international aid organizations to secure additional resources.

Healthcare and Managing Radiation Exposure

Governments will also need to provide healthcare to those affected by radiation exposure and other injuries sustained during the nuclear event. Radiation poisoning, burns, injuries from explosions, and psychological trauma will require significant medical attention.

Medical triage and treatment: Governments may set up emergency medical facilities to treat radiation poisoning and injuries. These facilities will prioritize those who are most severely affected, while also managing long-term health risks for survivors who have been exposed to lower levels of radiation.

Radiation testing and monitoring: Ongoing radiation testing and monitoring will be essential for managing public health. Governments may deploy mobile radiation testing units to check individuals for radiation exposure and determine whether they require treatment. Long-term monitoring will also be necessary for tracking the health effects of radiation over time.

Rebuilding Infrastructure and Communities

Once the initial danger has passed, governments will focus on rebuilding infrastructure that has been damaged or destroyed by the nuclear event. This includes restoring power grids, water systems, transportation networks, and communication systems.

Housing and shelter: Many survivors may have lost their homes or be unable to return due to radiation contamination. Governments will need to provide temporary housing and, eventually, assist in rebuilding or relocating communities.

Reconstruction efforts: Governments will play a central role in organizing the reconstruction of critical infrastructure. This will likely involve large-scale efforts to clear rubble, repair roads, restore water supplies, and rebuild power plants and other essential systems.

Long-Term Environmental Clean-up

One of the greatest challenges facing governments in the aftermath of a nuclear disaster is the long-term environmental clean-up required to make areas safe for human habitation again. Radioactive fallout can contaminate soil, water, and air, making it dangerous for years or even decades. Governments will need to spearhead efforts to decontaminate these areas.

Decontamination projects: Decontamination may involve removing topsoil, cleaning water sources, and using chemicals or biological agents to neutralize radiation. In heavily contaminated areas, entire regions may need to be evacuated and cordoned off for decades.

Radiation monitoring: Long-term radiation monitoring will be critical for determining which areas are safe for resettlement and which remain dangerous. Governments will likely establish "no-go" zones in highly radioactive areas, similar to the exclusion zone around Chernobyl.

Challenges Governments May Face

While governments will play a central role in post-nuclear survival, they will also face significant challenges. The scale of a nuclear disaster could overwhelm existing infrastructure and leadership, leaving governments struggling to meet the needs of their citizens.

Infrastructure Damage

Nuclear explosions could destroy key government facilities, military bases, hospitals, and communication centers. Without access to these critical resources, the government may find it difficult to coordinate a cohesive response. In such cases, local governments or military units may take on a greater role in managing survival efforts, while national governments focus on rebuilding.

Overwhelming Demand for Resources

A large-scale nuclear disaster would lead to massive shortages of food, water, medical supplies, and fuel. Governments will face immense pressure to meet the demands of a population desperate for resources. Without careful management, rationing, and cooperation with international aid organizations, resource shortages could lead to civil unrest and widespread suffering.

Communication Failures

In a post-nuclear world, communication networks may be down, making it difficult for governments to communicate with citizens. With radio towers, internet infrastructure, and phone lines damaged or destroyed, governments may have to rely on older, low-tech methods of communication, such as ham radio or physical message boards, to relay important information.

Limited Medical Capacity

Even before a nuclear event, most healthcare systems are not equipped to handle large-scale radiation exposure or mass casualties. After a nuclear disaster, hospitals and medical personnel may be overwhelmed, forcing governments to make difficult decisions about who receives treatment. This could lead to frustration and despair among those who cannot access necessary medical care.

What to Expect from Government Aid and How to Prepare

While governments will play an important role in managing post-nuclear survival, they may not be able to meet all of your needs. It's important to be prepared to survive on your own or within a community for extended periods. Here are some practical steps you can take to prepare for a post-nuclear world where government aid may be limited or delayed:

Stockpile Essentials

Stockpile food, water, medical supplies, and fuel before disaster strikes. This reduces your reliance on government resources and ensures you have enough to survive in the initial weeks or months when aid may be slow to arrive.

Build or Identify Shelters

If you live in an area that may be affected by nuclear fallout, identify or build a fallout shelter. Government shelters may be overcrowded or difficult to access in the aftermath of a disaster, so having your own shelter ensures that you and your family can stay safe from radiation.

Establish Communication Networks

While government broadcasts will be an important source of information, you should also establish your own communication networks. Ham radios, shortwave radios, and walkie-talkies can help you stay in contact with other survivors and receive updates when government communication systems fail. Learn how to operate these devices and make sure you have backups, such as hand-crank radios or solar chargers, to keep them running when power is unavailable.

Create a Self-Sustaining Community

While the government may provide assistance, long-term survival will require a degree of self-sufficiency. Forming a survival community with neighbors, friends, or fellow preppers can help distribute the burden of survival tasks. Together, you can pool resources, share skills, and defend against external threats.

Food production: Establish systems for growing your own food, such as small-scale gardening, hydroponics, or foraging. This reduces reliance on government food aid, which may be unpredictable or scarce.

Water collection and purification: Learn how to collect and purify water using methods like rainwater harvesting, filtration, or distillation. Governments may prioritize water distribution in more heavily populated areas, so having your own source of clean water is critical for long-term survival.

Skills exchange: Organize your community so that each person contributes based on their expertise. Someone with medical knowledge can help with healthcare, while others may be skilled in farming, security, or mechanics. Specialization within a community boosts overall survival chances.

Stay Informed but Be Cautious of Misinformation

While governments can be a reliable source of information, it's also important to be aware of the possibility of misinformation or confusion, especially in the chaos following a disaster. Governments may underplay radiation risks to prevent panic, or local authorities may not have the most up-to-date information. Use your own Geiger counters and dosimeters to independently verify radiation levels and rely on trusted sources within your community.

Use multiple sources: Try to get information from a variety of sources, including government broadcasts, ham radio operators, and community monitoring efforts. Cross-check this information to ensure you're making informed decisions based on the most accurate data available.

Be cautious of rumours: In a post-nuclear world, rumors can spread quickly, especially in areas where communication is limited. Before acting on any information, especially regarding radiation levels or government aid, try to confirm the details from multiple sources.

Have a Plan for When Government Aid Is Delayed or Unavailable

While government aid will be crucial, especially in the long term, there is always the risk that it may be delayed, insufficient, or fail to reach certain areas. It's important to have contingency plans for surviving without government assistance for extended periods.

Prepare for self-sufficiency: Plan to be self-sufficient for at least the first few weeks or months after a nuclear disaster. This means having enough food, water, and supplies to survive without relying on government rations.

Adapt to changing circumstances: Be flexible in your approach to survival. Government aid may not come in the form you expect, or it may be rationed unevenly. Be prepared to adjust your plans based on the availability of resources and the changing situation on the ground.

Potential Government-Led Initiatives for Long-Term Survival

If a functioning government remains after the initial phase of the nuclear disaster, it may implement long-term survival initiatives aimed at rebuilding society. These efforts will be essential for restoring infrastructure, providing healthcare, and addressing the environmental damage caused by nuclear fallout. Potential government-led initiatives could include:

Resettlement Programs

After decontamination efforts in certain areas, governments may organize resettlement programs to move survivors back into safe zones. This process could involve rebuilding housing, reestablishing basic services like electricity and water, and ensuring that radiation levels are low enough for safe habitation.

Safe zones: Governments will likely designate specific areas where radiation levels are sufficiently low for people to return. These areas may be prioritized for reconstruction efforts, and survivors will be encouraged to move there.

Assistance programs: Governments may provide materials, financial aid, or logistical support to help survivors rebuild homes and businesses in resettled areas.

Radiation Decontamination Projects

Long-term decontamination projects will be necessary to make large swaths of land habitable again. Governments may deploy specialized teams to remove radioactive materials from contaminated areas, particularly those needed for agriculture or housing.

Soil decontamination: Techniques such as removing topsoil, planting specific plants that can absorb radiation (known as phytoremediation), or using chemicals to bind radioactive particles may be employed to clean up contaminated land.

Water purification: Governments will likely prioritize decontaminating critical water sources, using advanced filtration techniques to ensure that clean water is available for survivors.

Economic Recovery Programs

Restoring the economy will be a significant challenge after a nuclear event. Governments may implement programs designed to restart industries, provide jobs, and encourage commerce. These programs could include subsidies for businesses, government contracts for rebuilding efforts, or job training for survivors.

Barter and trade systems: In the early stages of recovery, governments may encourage or regulate barter systems in areas where currency has lost its value. Communities might rely on trade to access resources they lack, and governments could facilitate these exchanges by setting up trade hubs or bartering markets.

International Aid and Collaboration

If international governments remain intact, global cooperation could play a crucial role in recovery. International aid organizations, neighboring countries, or multinational coalitions may send resources, medical supplies, and technical expertise to help with rebuilding efforts.

International relief: Governments may collaborate with international organizations such as the United Nations or Red Cross to provide food, medical care, and shelter to survivors. These efforts can supplement national programs that are struggling to meet the demand for aid.

Knowledge-sharing: International collaboration can also provide valuable knowledge for decontamination, healthcare, and rebuilding. Countries with expertise in nuclear disaster recovery may share best practices and technology with governments working to recover from a nuclear event.

The role of government in post-nuclear survival is multifaceted, ranging from immediate emergency response to long-term recovery and rebuilding. Governments are responsible for organizing evacuations, disseminating critical information, managing resources, and maintaining law and order. However, the scope and effectiveness of government aid will depend on the extent of the disaster and the resilience of governmental infrastructure. While government assistance will be vital, survivors should not rely solely on external aid. Whether government aid is delayed, insufficient, or unavailable, being proactive in managing your own survival will give you the best chance of weathering the crisis and helping rebuild society in the long term.

Staying Vigilant: Defending Against Desperate People

In the wake of a nuclear disaster, resources like food, water, and shelter will become scarce, creating a tense and potentially dangerous environment. As desperation rises, individuals and groups may resort to violence or theft to secure what they need to survive. This chapter focuses on the importance of staying vigilant and protecting yourself, your family, and your community from desperate people who may pose a threat in the post-nuclear world. We'll explore strategies for securing your home or shelter, building a defense plan, and managing interactions with outsiders in a way that minimizes risk while preserving your own survival.

Understanding the Threat: Why Desperation Leads to Danger

In the immediate aftermath of a nuclear disaster, most people will focus on survival—finding shelter, securing food and water, and staying safe from radiation. However, as time passes and resources become more scarce, the situation can devolve into chaos. People who may have been law-abiding citizens before the disaster could become desperate enough to engage in theft, violence, or other forms of aggression. It's essential to understand that the driving force behind this behavior is survival, and desperate people may act irrationally or aggressively if they believe their lives or the lives of their families are at stake.

The Breakdown of Social Order

When the normal structures of society collapse—law enforcement, emergency services, and government aid—people may feel that they have no choice but to take matters into their own hands. As resources dwindle, civil order may break down, leading to widespread looting, theft, and violence. In such a situation, protecting yourself and your resources becomes a top priority.

Competition for Resources

In a post-nuclear world, resources will be in high demand, and some individuals or groups may attempt to take what they need by force. This is particularly true if people believe that others have stockpiled food, water, medicine, or valuable supplies. Desperate individuals may see well-supplied groups or homes as targets, leading to potential conflicts over limited resources.

Organized Groups

While many threats will come from individuals acting out of desperation, some groups may organize for survival, and these groups could become aggressive or territorial. Organized bands of survivors may patrol areas, raid homes, or establish control over key resources like water sources or food supplies. Dealing with organized groups requires careful planning and, in some cases, negotiation or avoidance.

Securing Your Home or Shelter

The first line of defense against desperate people is securing your home or shelter. Whether you're in a house, bunker, or improvised shelter, taking steps to fortify your location can deter would-be intruders and give you a better chance of defending your resources.

Reinforce Entrances and Windows

The most common entry points for intruders are doors and windows, so reinforcing these areas is critical to securing your home.

Doors: Reinforce doors with deadbolt locks, metal bars, or barricades. If possible, use heavy-duty doors made of solid wood or metal that are harder to break down. Consider installing crossbars or bracing systems to prevent doors from being forced open.

Windows: Board up windows with plywood or metal sheeting to prevent them from being broken. If you still need natural light, use wire mesh or metal bars to cover the windows while allowing some visibility. Curtains or blackout material can also help keep your home hidden from view at night.

Secondary entrances: Don't forget to secure secondary entrances, such as back doors, basements, or garage doors. These entry points are often overlooked but can be vulnerable to intruders.

Establish a Perimeter

Securing the perimeter around your home or shelter is just as important as reinforcing the structure itself. A well-established perimeter can provide an early warning of approaching threats and deter intruders from getting too close.

Fencing and barriers: Set up physical barriers like fences, walls, or barricades around your property. Even improvised barriers like piles of debris, wooden stakes, or razor wire can slow down potential intruders and give you time to prepare.

Noise deterrents: Use noise-making devices, such as cans on strings, bells, or other makeshift alarms, to alert you to any movement outside your perimeter. These can be placed at entry points, gates, or along pathways that intruders might use.

Lighting: If you have access to electricity or solar-powered lights, use them strategically around your property to illuminate key areas at night. Motion-activated lights can startle intruders and provide early warning of their approach. However, be mindful of attracting attention with too much lighting, as it can signal that your location is occupied and well-stocked.

Create Safe Rooms

If intruders manage to breach your perimeter or enter your home, having a safe room can be the last line of defense. Safe rooms are fortified areas where you and your family can retreat to if your home is under attack.

Reinforced walls: Use thick, reinforced materials like concrete or metal to create a safe room that is resistant to forced entry.

Supplies: Keep essential supplies like food, water, and medical kits inside the safe room in case you need to shelter there for an extended period.

Communication and escape routes: Equip the safe room with a means of communication, such as a radio or cell phone, to call for help if possible. Additionally, plan escape routes in case you need to flee your shelter entirely.

Building a Defense Plan

In a world where desperate people may try to take what you have, it's essential to have a well-thought-out defense plan that includes strategies for securing your home, protecting your resources, and defending yourself if necessary. This plan should involve all members of your household or community and include clear protocols for dealing with potential threats.

Train for Self-Defense

Knowing how to defend yourself and your group is crucial in a post-nuclear world. While it's important to avoid violence whenever possible, there may be situations where you need to protect yourself from physical harm.

Self-defense techniques: Train in basic self-defense techniques that can help you disarm or incapacitate an attacker. Martial arts, close-combat skills, or even basic escape techniques can be life-saving in a crisis.

Weapons training: If you have access to firearms or other weapons, ensure that all responsible adults in your group are trained in their safe and effective use. Firearms can be a powerful deterrent, but they must be used responsibly to avoid accidents or unnecessary escalation.

Non-lethal options: In some cases, non-lethal methods such as pepper spray, stun guns, or batons may be preferable to firearms, especially if the goal is to deter intruders rather than engage in lethal combat.

Organize Group Defenses

If you're part of a survival community or have neighbors who are also focused on defense, organizing as a group can greatly increase your chances of successfully defending against intruders.

Rotating watch shifts: Set up a schedule for watch shifts so that someone is always keeping an eye on your perimeter. This ensures that your group remains vigilant, even when others are resting or attending to other tasks.

Communication plans: Establish a communication system within your group to quickly alert others in the event of a threat. This could be as simple as hand signals, whistles, or walkie-talkies.

Defensive positions: Identify key defensive positions around your home or community where group members can take cover or defend entry points. These positions should provide both visibility and protection from potential attackers.

Establish Rules for Dealing with Outsiders

In a post-nuclear world, not everyone you encounter will be a threat. Some people may seek help, join your group, or offer to trade resources. It's important to have clear rules for how your group deals with outsiders to avoid unnecessary risks.

Initial contact: Always approach outsiders cautiously. Keep them at a distance until you've assessed whether they pose a threat. Avoid bringing strangers into your shelter or showing them your resources until you've established trust.

Negotiation and trade: If you engage in trade or negotiation with outsiders, do so in a neutral location outside your perimeter. This minimizes the risk of them learning too much about your defenses or resources.

Warning and de-escalation: If someone becomes aggressive or tries to take what you have, your first course of action should be to warn them and try to de-escalate the situation. Make it clear that you are prepared to defend yourself but offer them an opportunity to leave peacefully.

Have an Evacuation Plan

If defending your shelter becomes impossible or the threat is too great, you may need to evacuate. Have an evacuation plan in place that includes alternative shelters or locations where you can retreat if necessary.

Pre-packed supplies: Keep "go bags" ready with essential supplies like food, water, medical kits, and warm clothing in case you need to leave your shelter quickly.

Escape routes: Identify multiple escape routes from your home or shelter. Know the safest paths to nearby areas where you can hide or regroup if your primary location becomes compromised.

Secondary shelters: Plan for secondary shelters in case you need to leave your main location. These could be remote locations, hidden bunkers, or temporary shelters that provide cover while you assess the situation.

Managing Mental and Emotional Vigilance

The stress of constantly being on guard can take a toll on your mental and emotional well-being. In a post-nuclear world, staying vigilant while managing anxiety and fear is critical for your overall health and survival. Here are some tips for maintaining mental resilience:

Balance Vigilance with Rest

While it's essential to stay alert, exhaustion can impair your ability to make sound decisions or react to threats. Ensure that everyone in your group gets adequate rest by rotating watch shifts and taking breaks when possible.

Build a Support Network

Lean on your community or group for emotional support. Share the burden of decision-making and defense planning so that no one feels overwhelmed by the responsibility of keeping everyone safe.

Practice Situational Awareness

Situational awareness—knowing what's happening around you at all times—is key to staying safe. Practice observing your environment, noting potential threats, and staying calm in stressful situations. Over time, this awareness will become second nature, allowing you to respond quickly and effectively to any dangers.

Defending yourself against desperate people in a post-nuclear world requires constant vigilance, preparation, and careful planning. By securing your home or shelter, building a defense plan, and organizing with your community, you can protect your resources and stay safe in an increasingly dangerous environment. Balancing self-defense with negotiation and de-escalation can help you avoid unnecessary conflict, but always be prepared to act decisively if your survival is at risk.

Food Preservation Techniques in a Fallout World

In a post-nuclear world, where food sources are scarce and traditional food distribution systems have broken down, preserving what little food you have becomes critical to survival. Radiation, contamination, and the collapse of modern infrastructure will make it difficult to grow fresh crops or hunt for uncontaminated meat, making food preservation techniques essential for maintaining a steady supply of nourishment. This chapter will explore the most practical and effective methods for preserving food in a fallout world, focusing on techniques that require minimal resources and rely on safe practices to avoid contamination and spoilage.

The Importance of Food Preservation in a Fallout World

Food preservation will play a key role in your survival after a nuclear disaster. Fresh food will be hard to come by, and any food that is gathered, grown, or scavenged must be preserved to last as long as possible. Without proper preservation methods, food will spoil quickly, leading to shortages that could be life-threatening in an environment where food production is severely limited.

Key reasons why food preservation is critical in a post-nuclear world:

Limited access to fresh food: Fallout contamination will make it dangerous to gather or grow fresh food in many areas. Preserving food allows you to build a stockpile and extend the life of safe, uncontaminated food sources.

Resource management: In a survival situation, it's vital to avoid waste. Preserving food ensures that nothing is lost to spoilage, allowing you to stretch your resources further.

Seasonal challenges: If you're able to grow crops or forage, certain foods may only be available seasonally. Preserving food allows you to save those items for times when they are not in season or when food supplies run low.

Contamination risks: In a world polluted by fallout, fresh food may carry radiation or other contaminants. Proper preservation methods can help remove or reduce these risks, especially when combined with thorough cleaning or cooking.

Key Food Preservation Methods in a Fallout World

Below are some of the most practical food preservation techniques that can be used in a fallout environment, focusing on methods that require minimal energy and resources while maximizing the longevity of your food supplies.

Drying and Dehydration

Drying is one of the oldest and simplest methods of preserving food. Removing moisture from food inhibits the growth of bacteria, mold, and yeast, allowing dried foods to last for months or even years. In a fallout world, where refrigeration may not be available, drying can be a highly effective way to preserve fruits, vegetables, meats, and grains.

Sun drying: In areas where fallout and radiation are not a major concern, sun drying can be used to dehydrate food. Lay thin slices of fruits, vegetables, or meat on a clean surface (such as a drying rack or cloth) in direct sunlight, ensuring proper air circulation. However, avoid this method in heavily contaminated areas, as fallout particles may settle on the food.

Air drying indoors: In a fallout environment, where contamination from the outside is a concern, indoor air drying is a safer option. Hang herbs, fruits, or meat in a well-ventilated room away from dust or fallout particles. You can create a simple drying rack using string or wire, and place food in thin layers to promote airflow.

Solar dehydrators: If you have access to materials, building a solar dehydrator can help dry food more efficiently while protecting it from fallout contamination. A solar dehydrator uses a glass or plastic enclosure to trap sunlight and dry food in a controlled environment. Plans for constructing a simple solar dehydrator can be found in survival guides or online resources.

Oven or stove drying: If you have access to a wood-burning stove or other heat sources, you can dehydrate food by placing it in the oven at a low temperature (around 120-140°F). This method requires close attention to avoid overcooking or burning the food.

Foods that can be dried: Fruits like apples, berries, and apricots; vegetables such as tomatoes, onions, and peppers; meats like jerky; and grains or beans can all be preserved using dehydration.

Smoking

Smoking is another traditional method of preserving food, particularly meat and fish. By exposing food to low heat and smoke, you can both dry the food and add antimicrobial properties from the smoke that help prevent spoilage.

Cold smoking vs. hot smoking: There are two main methods of smoking—cold smoking and hot smoking. Cold smoking (under 85°F) is a slower process that primarily dries the food, while hot smoking (between 160°F and 225°F) cooks the food while preserving it.

Building a smoker: If you don't already have a smoker, it's possible to build one using basic materials like wood, metal, or even bricks. A simple smoker can be made by digging a pit, lighting a small fire, and placing a rack above the fire to hold the food. The key is to control the smoke and heat to prevent overcooking or burning the food.

Types of wood: Different types of wood can be used to flavor the food during smoking. Hardwoods like hickory, oak, apple, and cherry produce excellent smoke for preserving meat. Avoid using softwoods like pine, which can impart a bitter taste and produce harmful toxins.

Smokable foods: Meats like fish, poultry, beef, and pork are ideal for smoking. Smoking can also be used to preserve cheese or certain vegetables, though this is less common.

Salting and Curing

Salting is an effective way to preserve meat and fish by drawing moisture out of the food and creating an environment that inhibits the growth of bacteria. In the absence of refrigeration, salting can extend the shelf life of meat for several months.

Dry salting: Dry salting involves rubbing coarse salt directly onto the surface of meat or fish, ensuring that the entire piece is coated. Once salted, the meat should be stored in a cool, dry place for several days or weeks, depending on the size of the cut. As the salt absorbs moisture, it creates a hostile environment for bacteria.

Brining: Brining is a similar method that uses a saltwater solution to preserve food. Immerse meat, fish, or vegetables in a saltwater brine (typically 1 cup of salt per gallon of water) and store the container in a cool, dark place. The brine will help draw moisture from the food while protecting it from spoilage.

Salt-cured foods: Fish, pork, beef, and even certain vegetables can be salt-cured. Traditional dishes like salt cod or ham have been preserved using this method for centuries, and salt curing remains a reliable way to store food without refrigeration.

Fermentation

Fermentation is a natural preservation method that uses beneficial bacteria to convert sugars in food into acids, alcohol, or gases. These by-products help preserve the food while adding nutrients and flavors.

Lacto-fermentation: Lacto-fermentation is the process of using salt to create an environment where lactic acid bacteria thrive. This method is commonly used to preserve vegetables like cabbage (sauerkraut), cucumbers (pickles), and peppers. The salt not only inhibits harmful bacteria but also allows the fermentation process to begin.

How to ferment food: To ferment vegetables, pack them tightly into a container (such as a glass jar) and add salt or a saltwater brine. The vegetables should be fully submerged in the brine to prevent mold growth. Store the container in a cool, dark place, and allow the fermentation process to take place over the course of several days to weeks, depending on the desired flavor and texture.

Alcohol fermentation: Alcohol fermentation can be used to preserve fruits, particularly by making fruit-based alcohols like cider or wine. In a fallout world, this may be a useful method for both preserving fruit and producing an alternative source of clean, safe liquid.

Fermented foods: Sauerkraut, kimchi, pickles, and yogurt are all examples of foods preserved through fermentation. Fermented foods can last for months, making them an important source of vitamins and nutrients during food shortages.

Canning

Canning involves preserving food in airtight containers that are heated to a high temperature to kill bacteria and other microorganisms. Properly canned food can last for years if stored correctly, making it one of the most reliable long-term preservation methods.

Water bath canning: Water bath canning is best suited for high-acid foods like fruits, tomatoes, and pickles. To can using this method, fill jars with food, seal them with lids, and submerge them in boiling water for a set amount of time (usually 10–30 minutes). The heat kills harmful bacteria, and the vacuum seal prevents new contamination.

Pressure canning: Pressure canning is required for low-acid foods like meats, vegetables, and soups. Using a pressure canner, which reaches temperatures higher than boiling water, is essential to prevent botulism and other harmful bacteria from surviving the canning process. Pressure canners are more complex than water bath canners but are essential for preserving protein-rich foods.

Canned foods: Meats, stews, fruits, vegetables, and even jams can be preserved using canning methods. In a fallout scenario, canning is ideal for stockpiling large quantities of food that need to be stored for an extended period.

Root Cellaring

Root cellaring is a low-tech, low-energy method of food preservation that relies on cool, stable temperatures and humidity levels to keep food fresh for months. While it doesn't extend the shelf life of food as long as other methods, root cellaring can be an excellent way to store root vegetables and hardy crops during the colder months.

Building a root cellar: A root cellar can be dug into the ground, using the natural insulation of the earth to maintain a cool, consistent temperature. Alternatively, you can convert a basement or underground storage area into a makeshift root cellar by adding ventilation and shelves for storing produce.

Foods stored in root cellars: Potatoes, carrots, onions, garlic, beets, and other root vegetables are ideal for storing in a root cellar. Apples and winter squash can also be stored in this way, provided the cellar is cool and ventilated.

Managing Contamination Risks in a Fallout World

In a world polluted by nuclear fallout, contamination is a constant threat, and preserving food safely requires extra precautions to prevent exposure to radioactive particles or other harmful substances.

Thorough washing: Before preserving any food, thoroughly wash it with clean, uncontaminated water to remove any fallout particles that may have settled on the surface. Scrub vegetables and fruits, and rinse grains or beans to reduce the risk of contamination.

Peeling and trimming: For root vegetables or fruits, peeling the outer layer can help remove surface contamination. Additionally, trimming the tops and ends of vegetables may reduce the amount of radiation-absorbed material.

Filtering water: If you're using water for preservation (such as in canning or brining), ensure that the water is purified or filtered to remove radioactive contaminants. Boiling, distillation, or the use of a high-quality water filter can help ensure that the water used in preservation is safe.

In a post-nuclear world, preserving food will be one of the most important skills for survival. With the right techniques—drying, smoking, salting, fermenting, canning, and root cellaring—you can extend the life of your food supply and build a stockpile that will sustain you through the harshest times. While contamination risks are always present in a fallout environment, taking the proper precautions can help minimize exposure and keep your preserved food safe for consumption. By mastering these preservation methods, you can increase your chances of survival in a world where fresh food is scarce and every bite matters.

Safe Zones: Where Can You Go After a Nuclear War?

In the aftermath of a nuclear war, finding a safe place to live becomes a critical task. Radiation fallout, contaminated environments, and the destruction of infrastructure make large swathes of land uninhabitable. Safe zones—areas where radiation levels are low enough for human survival—become refuges for survivors seeking to avoid the dangers of radiation exposure, scarcity of resources, and societal breakdown. This chapter explores where you might be able to go after a nuclear war, what to look for in a safe zone, and how to evaluate whether an area is suitable for long-term habitation.

What is a Safe Zone?

A safe zone is an area where radiation levels have dropped to survivable levels, resources like food and water are accessible, and the environment is stable enough to support human life. These areas are not necessarily free from all dangers but provide a relatively safer option compared to the highly irradiated and devastated regions following a nuclear conflict.

Characteristics of a Safe Zone:

Low radiation levels: Safe zones have radiation levels low enough to minimize long-term health risks. This means exposure is within limits that are manageable for human survival.

Stable environment: Safe zones have access to relatively uncontaminated water sources, fertile land for growing food, and shelter from environmental hazards like fallout storms.

Sufficient resources: Access to food, water, and medical supplies is critical. Safe zones should have either natural resources or infrastructure that can be repaired or maintained for survival.

Security: Safety from both environmental dangers and other desperate survivors is essential. Safe zones should be defensible and free from widespread violence or lawlessness.

Determining Safe Zones After a Nuclear War

Identifying a safe zone after a nuclear war involves evaluating several factors, including radiation levels, geographical features, proximity to fallout zones, and access to essential resources. Knowing where radiation is likely to be lower and how to test areas for safety will help you determine where to go.

Distance from the Blast Zones

The further you are from nuclear blast sites, the lower the risk of dangerous radiation levels. Areas close to the epicentres of nuclear detonations will experience the highest levels of radiation, especially in the immediate aftermath. These zones will remain uninhabitable for years or even decades.

Primary targets: Major cities, military installations, and key industrial areas are likely to be primary targets in a nuclear war. Safe zones should be located far from these areas to minimize exposure to high levels of fallout.

Downwind danger: Fallout spreads downwind from blast zones, carried by prevailing winds. Areas directly downwind of a nuclear detonation will experience higher radiation levels as fallout particles settle. Avoid regions in the direct path of prevailing winds from known blast sites.

Testing radiation levels: Use a Geiger counter or dosimeter to test the radiation levels in any area you consider a potential safe zone. Radiation levels below 0.1 μSv/h are generally safe for long-term habitation, while anything above 1 μSv/h can pose serious risks.

Geographical Features and Natural Barriers

Geography plays a crucial role in determining whether an area can serve as a safe zone. Natural barriers like mountains, hills, and forests can protect areas from fallout and provide shelter from the elements.

Mountains: Mountain ranges can act as natural shields against radiation fallout carried by winds. High-altitude areas are less likely to accumulate heavy fallout, making them potential safe zones. The terrain also provides natural fortifications against human threats.

Valleys and basins: Areas at lower elevations, such as valleys or basins, may be less affected by direct fallout but can accumulate radioactive dust over time due to natural wind patterns. Evaluate these areas carefully and test radiation levels regularly.

Forests: Dense forests may help reduce exposure to fallout by acting as a barrier, trapping radioactive particles in the trees and soil. However, forests near blast zones can become heavily contaminated, so they are only suitable if far from detonation sites.

Water sources: Clean, uncontaminated water is critical for any safe zone. Choose areas near fresh water sources like rivers, lakes, or natural springs that are located far from fallout zones. Test the water regularly for radiation contamination to ensure it's safe for consumption.

Remote and Less Populated Areas

Remote areas, particularly those with low population density, are more likely to remain untouched by nuclear strikes and fallout. These regions may offer the best chances for finding a safe zone after a nuclear war.

Rural and wilderness areas: Rural areas, particularly those far from major cities or industrial centers, are less likely to be targeted in a nuclear attack. These areas may have lower radiation levels and offer more opportunities for hunting, foraging, or farming.

Islands: Islands located far from the mainland and major urban centers may avoid the worst of the fallout. Remote islands could serve as ideal safe zones, provided they have access to clean water, arable land, and some form of renewable resources like fish or sea plants.

National parks and reserves: Nature reserves, parks, and wilderness areas may provide safe havens, especially if they are far from blast zones. These areas often have fewer people and better access to natural resources, making them suitable for long-term survival.

Sheltering Underground or in Natural Caves

One of the safest ways to avoid radiation in the short term is by taking shelter underground. Fallout radiation is most dangerous when you are exposed to it directly, and underground shelters can provide protection while radiation levels decrease over time.

Natural caves: Caves can serve as excellent temporary or permanent shelters in a fallout world. The thick rock provides natural shielding from radiation, and caves often maintain a stable internal temperature, making them comfortable for long-term habitation.

Underground bunkers: If you can find or build an underground bunker, this is one of the best ways to stay safe from radiation. Bunkers should be reinforced and equipped with proper ventilation and supplies for long-term survival. These shelters offer excellent protection from fallout and are a top choice for those who need to remain in high-risk areas.

International Safe Zones

Depending on the scale of the nuclear conflict, certain countries or regions may be less affected by the fallout and offer safer environments for survivors. Some international safe zones might be neutral countries or remote regions that were not involved in the conflict.

Southern Hemisphere: Historically, much of the nuclear arsenal is concentrated in the Northern Hemisphere. Countries in the Southern Hemisphere—such as Australia, New Zealand, and parts of South America—are likely to experience less direct fallout. These countries may become havens for those able to travel long distances.

Isolated regions: Countries or territories that are geographically isolated—such as Iceland, Greenland, or parts of the Arctic—may avoid much of the fallout and remain safer than heavily populated areas.

International relief zones: After a nuclear conflict, international organizations may establish safe zones in less affected regions. These relief zones may be designed to accommodate refugees from heavily irradiated areas and provide access to clean water, food, and medical care.

Evaluating a Potential Safe Zone

Before settling in a new area, it's important to thoroughly evaluate whether the location is truly safe for long-term survival. This involves not only testing for radiation but also assessing the availability of essential resources and the overall sustainability of the environment.

Testing Radiation Levels

Using a Geiger counter or dosimeter, regularly test the area for radiation levels. Take multiple readings at different times of day and in different parts of the area to ensure the zone is consistently safe. Avoid any area where readings are consistently high, as prolonged exposure to radiation can cause serious health problems.

Evaluating Water and Food Sources

Access to clean water is the most important factor when determining whether an area is suitable as a safe zone. Ensure that there are reliable sources of fresh water, such as rivers, lakes, springs, or wells. Test water for contamination, as fallout particles can easily enter water supplies.

In terms of food, check whether the area has fertile soil for growing crops or opportunities for hunting and fishing. If foraging, be cautious of contamination and thoroughly clean and inspect any food before consuming it.

Safety from Human Threats

In addition to environmental dangers, safe zones must also offer protection from other survivors who may be desperate or violent. Ensure that your chosen area is defensible, whether through natural barriers, distance from heavily populated regions, or the availability of weapons and supplies for self-defense.

Moving to a Safe Zone: Timing and Strategy

Knowing when and how to move to a safe zone is critical for minimizing exposure to radiation and other dangers. In the immediate aftermath of a nuclear war, it's best to shelter in place for at least the first 24 to 48 hours while radiation levels are at their highest. Once radiation levels decrease, you can begin planning your journey to a safer area.

Shelter in Place First

For the first few days following a nuclear detonation, fallout levels will be extremely high. Stay indoors, in a basement, or in another protected area until the initial fallout settles and radiation levels drop to safer levels. Moving too soon can expose you to lethal doses of radiation.

Plan Your Route

If you need to relocate to a safe zone, plan your route carefully to avoid heavily contaminated areas. Use radiation monitoring tools along your journey to test each area you pass through. Travel through less populated, rural areas, as these are less likely to have been targeted in the initial strikes.

Travel Light but Prepared

Pack only the essentials for your journey: food, water, a Geiger counter, a map, warm clothing, and first aid supplies. Travel light to move quickly and efficiently, but ensure you have enough supplies to survive the trip.

In a post-nuclear world, finding and moving to a safe zone is critical for your long-term survival. By understanding radiation patterns, evaluating geographical features, and testing for contamination, you can identify areas that are more likely to be safe from the immediate and long-term effects of nuclear fallout. While the road to a safe zone may be dangerous, careful planning, resource management, and strategic thinking will increase your chances of finding a place where you and your group can rebuild and thrive.

Building a Livestock Pen in Fallout Zones

In a post-nuclear world, livestock can be a critical resource for providing food, labor, and even companionship. However, raising animals in a fallout zone comes with significant challenges. Radioactive contamination in the environment can pose serious health risks to both livestock and humans, and managing animals in a safe, controlled manner becomes essential. This chapter explores how to build a livestock pen in a fallout zone, including the strategies for protecting animals from radiation exposure, securing their food and water supply, and ensuring their health and safety. By implementing these practices, you can maintain a sustainable livestock system that helps support your survival in the long term.

The Importance of Livestock in a Fallout World

In the aftermath of a nuclear war, food sources become scarce, and traditional farming may be difficult due to contaminated soil and water. Livestock offers a valuable alternative for producing food and other resources, such as milk, eggs, and wool. Additionally, animals can provide labor, such as hauling supplies, and manure for fertilizing crops.

Key benefits of raising livestock in a fallout world:

Food supply: Livestock can provide a renewable source of protein, including meat, milk, and eggs, helping to sustain your group in times of food scarcity.

Fertilizer: Manure from livestock can be used to fertilize soil, especially in areas where growing crops is possible, helping to regenerate agriculture.

Labor and transportation: Larger animals, such as goats or donkeys, can assist with tasks like pulling carts or carrying supplies.

Sustainability: Livestock can help create a self-sustaining survival system, reducing reliance on external food sources.

However, to reap these benefits, livestock must be kept healthy and protected from the radiation and contamination present in the fallout environment.

Choosing Livestock for Fallout Zones

Not all animals are suitable for survival in a post-nuclear world. Some species are more resilient to environmental stress and easier to manage in difficult conditions. When selecting livestock, prioritize animals that can survive on minimal resources, are hardy in harsh environments, and reproduce quickly.

Hardy Livestock Species

Some livestock species are more adaptable and resistant to the challenges posed by a fallout zone. These animals tend to require less intensive care and can survive in environments where food and water are limited.

Chickens: Chickens are highly versatile and resilient. They require relatively little space, produce eggs regularly, and can survive on a variety of feed sources, including foraged plants, insects, and food scraps. They also reproduce quickly, making them a sustainable food source.

Goats: Goats are known for their hardiness and ability to thrive in a wide range of environments. They provide milk, meat, and can even be used for labor. Goats are excellent foragers and can survive on tough vegetation that other animals might avoid.

Sheep: Sheep are another good option for fallout zones. They provide meat, wool for clothing, and manure for crops. Sheep are also relatively easy to manage in terms of feeding and shelter requirements.

Pigs: Pigs are efficient at converting food waste into meat, making them a valuable resource in a survival scenario. They can also be fed foraged plants and leftover food, though they require careful management to avoid contamination.

Rabbits: Rabbits are easy to breed, grow quickly, and require little space, making them ideal for small-scale survival systems. Their manure is an excellent fertilizer for crops, and they can be fed simple forage such as grasses or leafy greens.

Livestock Suitability in Fallout Zones

The key challenge for livestock in a fallout zone is exposure to radiation, either through direct contact with radioactive particles or by consuming contaminated food and water. Animals raised in these conditions will require special care to prevent radiation sickness and ensure their long-term health.

Radiation resistance: Some animals, like chickens and goats, are more resistant to radiation exposure than others. While no animal is immune to radiation, species that are smaller, reproduce quickly, and can forage for food are more likely to survive in contaminated environments.

Access to shelter: Animals that can be housed in covered or indoor environments, such as barns or shelters, are better protected from fallout. Large animals that must spend significant time outdoors may be at greater risk of contamination.

Building a Livestock Pen in a Fallout Zone

When constructing a livestock pen in a fallout environment, it's essential to protect your animals from radiation, contamination, and predators. The pen should be designed to minimize exposure to fallout while providing adequate space, ventilation, and access to food and water. Below are key steps for building a livestock pen that will help your animals thrive in a challenging environment.

Choosing a Location

The location of your livestock pen is critical to minimizing exposure to fallout and other environmental hazards. When selecting a site, consider factors such as proximity to contaminated areas, natural barriers, and shelter from the elements.

Avoid fallout hotspots: Place your livestock pen as far as possible from known blast sites or areas with high radiation levels. Avoid positioning the pen in low-lying areas where fallout particles may accumulate or where runoff from contaminated areas could reach.

Use natural barriers: If available, locate the pen near natural barriers such as hills, dense forests, or rock formations that can help shield animals from radioactive dust or windborne fallout. These barriers can also protect against strong winds or storms.

Shelter from the elements: Choose a location that offers protection from harsh weather conditions. Radiation aside, livestock still needs protection from the wind, rain, and extreme temperatures. Natural shelters or areas near existing structures can help provide this protection.

Building the Pen Structure

The pen itself should provide a balance of protection from radiation while allowing adequate ventilation and light for the animals. You can use a variety of materials to construct a pen, depending on what is available.

Radiation shielding: To minimize radiation exposure, the pen should include elements that block or reduce the amount of fallout that reaches the animals. Use materials like concrete, metal sheeting, or wood to create walls or roofs that shield the animals from radioactive dust. If concrete is unavailable, thick layers of wood, earth, or even sandbags can provide effective shielding.

Covered enclosures: Build covered enclosures where the animals can be kept during periods of heavy fallout or high radiation levels. These shelters should be completely enclosed and equipped with doors or flaps to prevent fallout from entering. When fallout is not as severe, animals can be allowed to roam in enclosed outdoor areas with some protection.

Ventilation and drainage: Ensure that the pen has proper ventilation to prevent the buildup of harmful gases, moisture, and heat inside the structure. However, be mindful of leaving openings where fallout could settle. Install screens or filters over vents to block dust. Additionally, make sure the pen has adequate drainage to prevent water from pooling, as contaminated water can be a source of radiation.

Protecting Food and Water Supplies

In a fallout zone, the food and water supply for your livestock must be carefully managed to prevent contamination. Contaminated feed or water can quickly lead to radiation sickness and death in animals, so proper precautions are necessary.

Water filtration: If the available water supply is potentially contaminated, use filtration systems to remove radioactive particles from the water before giving it to livestock. Boiling water can remove bacteria, but it does not remove radioactive materials. Instead, use specialized water filters designed to eliminate heavy metals and radioactive particles.

Covered feed storage: Store animal feed in airtight, covered containers to protect it from fallout dust and moisture. Avoid feeding animals from open fields or pastures where fallout may have settled. Instead, provide foraged food from uncontaminated sources or grow feed indoors where it is shielded from radiation.

Foraging and grazing considerations: If grazing is necessary, limit the amount of time animals spend foraging in exposed areas. Instead, rotate grazing areas and consider planting crops or grasses in indoor environments like greenhouses or enclosed gardens where radiation exposure is reduced.

Managing Animal Health

Maintaining the health of your livestock in a fallout zone is a continuous process. This requires regular monitoring for signs of radiation exposure and taking proactive steps to reduce the risks.

Daily health checks: Monitor your animals daily for any signs of illness, particularly symptoms of radiation exposure, such as lethargy, loss of appetite, or hair loss. Early detection is crucial for treating affected animals or culling those that are too sick to recover.

Decontamination: If your animals are exposed to fallout, decontaminate them by washing their fur, feathers, or skin with clean, uncontaminated water. Use a mild soap to remove dust or particles that may have settled on their bodies. Pay special attention to their mouths, eyes, and nostrils, where contamination can easily occur.

Medical care: Stockpile basic veterinary supplies to treat injuries or illnesses in your animals. Antibiotics, antiseptics, and bandages are essential for managing common ailments. If possible, learn how to perform basic veterinary care, such as treating wounds or administering medicine.

Long-Term Considerations for Raising Livestock

While raising livestock in a fallout zone presents numerous challenges, with careful planning, it can become a sustainable way to provide food and resources for survival. Here are some long-term considerations for ensuring the success of your livestock system:

Breeding and Population Management

In the long term, managing the breeding and reproduction of your livestock is critical for sustaining a stable population. Select the healthiest animals for breeding, focusing on those that show resilience to environmental stress and disease. Avoid inbreeding by maintaining genetic diversity, particularly if your livestock population is small.

Rotating Pastures and Feeding Areas

To minimize radiation exposure, rotate your livestock between different grazing areas. This prevents animals from over-foraging in contaminated areas and reduces the buildup of radioactive particles in the soil. If possible, create enclosed or sheltered feeding zones to further protect animals from contamination.

Supplementing Feed with Indoor Crops

Growing feed indoors, in greenhouses or enclosed environments, is a viable long-term solution for providing livestock with safe food. Crops like grass, alfalfa, or grains can be grown in controlled conditions, protecting them from fallout contamination. This system also helps reduce reliance on outdoor grazing and foraging, which may be compromised due to radiation.

Building a livestock pen in a fallout zone is a challenging but achievable task, one that can significantly enhance your chances of survival by providing a renewable food source and valuable resources. By carefully selecting hardy livestock, constructing a well-shielded pen, and protecting food and water supplies from contamination, you can maintain a healthy livestock population in even the harshest environments. Regular health monitoring, decontamination practices, and long-term breeding strategies will ensure that your animals remain productive and sustainable, contributing to the overall resilience of your survival plan. With the right precautions, livestock can become a key pillar of your strategy for thriving in a post-nuclear world.

Rationing Food and Water for Long-Term Survival

In a post-nuclear world, food and water become invaluable resources. The collapse of supply chains, contamination from radiation, and the destruction of infrastructure make it difficult to find fresh supplies. Rationing your available food and water carefully is essential for ensuring long-term survival. By managing these resources wisely, you can extend your supply, reduce the risk of starvation or dehydration, and maintain enough sustenance to support you and your group until new sources of food and water become available. This chapter explores effective strategies for rationing food and water, determining daily needs, and creating a plan that balances nutrition with conservation.

Why Rationing is Essential in a Fallout Environment

In a survival situation, especially after a nuclear disaster, the scarcity of food and clean water can quickly become a life-threatening issue. Fallout can contaminate crops, water sources, and animals, making many traditional sources of sustenance unavailable or dangerous to consume. Rationing ensures that your available supplies last as long as possible, helping you stretch resources through the immediate crisis and into the longer recovery period.

Key reasons why rationing is critical:

Limited resupply: In the early stages after a nuclear event, there may be no way to restock food or water supplies. Rationing ensures you don't run out of vital resources too quickly.

Managing contamination risks: Safe food and water will be hard to come by, especially with the risk of radioactive contamination. Rationing allows you to rely on known, uncontaminated supplies for as long as possible.

Preventing waste: Without rationing, it's easy to consume more than necessary, leading to shortages and potential starvation later on.

Assessing Your Food and Water Supply

Before creating a rationing plan, it's important to take inventory of your current food and water supplies. This will help you understand how much you have available and how long it can last when rationed properly.

Food Inventory

Take stock of all available food, including canned goods, dried foods, preserved meats, grains, and any fresh produce that may still be safe to eat. Categorize your food by shelf life and nutritional value, prioritizing items that will expire soon or provide essential nutrients.

Long shelf-life foods: Canned goods, dried beans, rice, pasta, and freeze-dried meals should be your primary food sources. These items are calorie-dense and can last for months or even years if stored properly.

Fresh foods: If you have access to fresh fruits, vegetables, or meat, these should be consumed first, as they will spoil the fastest. After a nuclear event, fresh food will be difficult to come by, so make the most of what you have before it goes bad.

Foraged or scavenged food: If you are able to forage or hunt, include these food sources in your inventory, but always be cautious of contamination. Thoroughly clean and cook any wild food to reduce the risk of radiation or other toxins.

Water Inventory

Water is even more critical than food for long-term survival. Humans can only survive a few days without water, and finding clean, uncontaminated water in a fallout environment can be challenging. Take stock of all available water, including bottled water, filtered water, or water collected from rain or other sources.

Stored water: If you've prepared in advance, you may have several gallons of stored water. Count how many gallons you have and ensure the containers are sealed and uncontaminated.

Natural water sources: If you live near rivers, lakes, or other natural water sources, these can supplement your supply. However, you'll need a reliable way to test and purify this water before consuming it, as fallout can easily contaminate natural bodies of water.

Water filtration systems: If you have water filters or purification tablets, include these in your inventory. These tools will allow you to make contaminated water safe to drink, increasing the number of viable water sources.

Determining Daily Food and Water Needs

Once you have assessed your inventory, the next step is determining how much food and water each person in your group will need per day. Rationing requires finding a balance between providing enough sustenance for survival while conserving resources for as long as possible.

Daily Caloric Needs

In a survival situation, your body will need enough calories to maintain energy levels, but you may need to reduce intake compared to normal living conditions. The average adult requires between 2,000 and 2,500 calories per day under normal conditions. However, in a survival scenario, you can reduce this intake to conserve food, though it's important to avoid malnutrition.

Men: Aim for 1,800 to 2,000 calories per day if physically active, or 1,500 to 1,800 calories if sedentary.

Women: Aim for 1,600 to 1,800 calories per day if physically active, or 1,200 to 1,600 calories if sedentary.

Children: Children will require fewer calories, typically between 1,000 to 2,000 calories depending on age and activity level.

Balancing calories and nutrients: Focus on calorie-dense foods like rice, beans, and fats, but don't neglect nutrients like vitamins and protein. A balanced diet helps maintain energy and immune system function.

Water Requirements

Each person requires at least 1 gallon of water per day for drinking and basic hygiene. In a rationing scenario, you may need to reduce this amount, but drinking less than half a gallon per day can quickly lead to dehydration. If water is extremely scarce, prioritize drinking water over other uses, such as hygiene.

Adult water needs: 1 gallon per day is ideal, but as little as 0.5 gallons per day can suffice in extreme conditions for drinking only.

Children's water needs: Children will need less water, typically around 0.5 to 1 gallon per day, depending on age and activity level.

Heat and activity: If your environment is hot or if you are engaging in strenuous activity (such as gathering supplies or building shelters), water needs will increase.

Creating a Rationing Plan

A rationing plan outlines how much food and water will be consumed daily, how supplies will be distributed, and how you'll handle special circumstances such as sickness or increased physical activity. The goal is to ensure that your supplies last as long as possible while providing enough sustenance to keep you and your group healthy.

Rationing Food

When rationing food, divide your daily caloric needs among the available food sources, ensuring that everyone receives an adequate mix of calories, protein, fats, and vitamins. Here's how to set up a basic food rationing plan:

Divide meals: Plan for two or three small meals per day rather than one large meal. This helps prevent hunger and keeps energy levels more stable throughout the day.

Focus on non-perishables: Rely on non-perishable foods like rice, beans, pasta, and canned goods as the backbone of your diet. Use fresh foods first but conserve non-perishables for long-term use.

Protein sources: Ensure everyone gets some form of protein each day, such as beans, canned meat, or preserved fish. Protein helps maintain muscle and immune system function.

Supplements: If available, consider rationing multivitamins or other nutritional supplements to prevent deficiencies.

Rationing Water

Water is critical to survival, so your rationing plan should carefully account for how much water is consumed each day. Here's how to approach water rationing:

Daily distribution: Assign each person a set amount of water per day based on your available supply. This may range from half a gallon to 1 gallon, depending on the situation.

Prioritize drinking: Drinking water takes priority over other uses. Hygiene, washing, and cooking can be managed with smaller amounts of water if necessary.

Conserving water: Reduce water use by practicing efficient hygiene (such as sponge baths instead of full washing) and reusing water for multiple purposes (such as using cooking water for cleaning).

Special Considerations

Certain circumstances may require adjustments to your rationing plan. For example:

Illness: If someone becomes ill, they may require more water to stay hydrated or a special diet to recover. Plan for extra rations in case of medical emergencies.

Increased activity: If you are physically active (e.g., building shelters, gathering supplies), you may need to increase your calorie and water intake slightly to maintain energy levels.

Children and vulnerable individuals: Adjust rations for children, the elderly, or those with specific medical conditions to ensure their nutritional needs are met.

Managing and Adapting Your Rationing Plan

Rationing is an ongoing process that requires regular monitoring and adjustment. As time passes, your available resources will change, and new sources of food or water may become available through foraging, scavenging, or trade.

Track Consumption

Keep a daily log of how much food and water is consumed by each person. This helps you monitor your supplies and identify when you may need to adjust rations or look for alternative food sources.

Inventory checks: Regularly check your inventory to see how much food and water remains. Calculate how long it will last at your current rationing levels, and adjust the plan if necessary.

Find New Sources of Food and Water

As your supplies dwindle, it's important to seek out new sources of food and water. This might include:

Foraging: In areas not heavily affected by fallout, foraging for wild plants, berries, nuts, or edible roots can help supplement your food supply. Be cautious of contamination, and always test or clean food thoroughly before eating.

Hunting or fishing: If you have the skills and equipment, hunting small game or fishing can provide additional protein. However, be sure to check for signs of radiation contamination in any animals you harvest.

Water collection: Set up rainwater collection systems if possible. Rainwater can be filtered and purified for safe drinking, providing a renewable water source.

Adjust Rations as Needed

As your situation changes, be prepared to adjust your rationing plan. If your food supply is running low, consider reducing daily caloric intake or cutting back on less essential foods. If you find new food sources, increase rations to keep everyone healthy and energized.

The key to successful rationing is balance—providing enough sustenance to maintain health while conserving supplies to ensure survival in the long term. With careful planning and adaptation, you can navigate the challenges of food and water shortages and emerge stronger in the face of adversity.

Rebuilding Society: A Future after Fallout

Rebuilding society after a nuclear fallout may seem like a daunting and distant goal, but it's a critical consideration for long-term survival and recovery. The immediate aftermath of a nuclear war or disaster will be chaotic, with widespread destruction, radiation, and resource scarcity dominating the landscape. However, as the radiation decays and survivors begin to organize, the focus will shift from basic survival to the larger task of rebuilding communities, infrastructure, and, eventually, entire societies.

This chapter will explore what it takes to rebuild society after a nuclear fallout, from establishing small communities and restoring essential services to re-establishing governance, trade, and long-term sustainability. The goal is to understand how humanity can move beyond survival and begin the process of creating a stable and functioning society again.

The Immediate Aftermath: Survival to Stabilization

In the days, weeks, and months after a nuclear disaster, survival is the primary focus. As the dust begins to settle and survivors start forming communities, the transition from mere survival to long-term stability can begin. The first stage in rebuilding society is establishing small, functioning groups of people who can work together for mutual benefit.

Forming Survivor Communities

Survivors will likely begin to form small groups or communities out of necessity. These groups may be made up of families, friends, neighbors, or strangers brought together by circumstance. Establishing a cooperative community is essential for pooling resources, sharing skills, and defending against threats.

Leadership and decision-making: Early on, communities will need to establish leadership or some form of governance to make decisions about resource distribution, defense, and long-term planning. Leadership doesn't have to be authoritarian—it can be democratic, with decisions made collectively, or based on expertise and skills.

Pooling resources: In the immediate aftermath, pooling resources—such as food, water, medical supplies, and tools—ensures that the entire group benefits and increases the chances of survival. A community that shares and cooperates will fare better than individuals hoarding supplies.

Security and defense: Small communities will need to defend themselves against external threats, including other desperate survivors, looters, and even wildlife. Organizing security patrols, setting up defenses, and training in self-defense are all critical tasks in this early phase.

Restoring Basic Services

Once a stable community is established, the next priority will be restoring basic services such as clean water, food production, shelter, and medical care. In the absence of modern infrastructure, survivors will have to rely on primitive methods and innovative solutions.

Water: Clean water is essential for survival. Communities will need to establish water collection and purification systems, using rainwater, rivers, lakes, or wells. Filtration and purification methods will be crucial to avoid contamination from fallout.

Food: Communities will need to focus on growing food in safe, uncontaminated areas. This might involve planting crops in enclosed greenhouses or starting small-scale farming with animals like chickens or goats. Foraging, hunting, and fishing can supplement food supplies but must be done cautiously to avoid contaminated areas.

Shelter: In many cases, survivors may be living in damaged homes or makeshift shelters. Communities will need to repair and fortify structures to provide safe, long-term housing. Building underground bunkers or using natural caves may be necessary in areas where radiation remains a threat.

Medical care: Medical resources will be limited, so survivors must manage injuries, illness, and radiation sickness with minimal supplies. Communities may have to rely on basic first aid, herbal medicine, and the skills of any surviving medical professionals.

Rebuilding Infrastructure: Laying the Foundations for Society

As the immediate survival phase ends and communities stabilize, the focus will shift toward rebuilding infrastructure and establishing systems that can support a larger society. This phase involves restoring communication, transportation, and energy sources, as well as developing a sustainable economy based on trade and production.

Restoring Communication

In a post-nuclear world, communication systems like the internet, phones, and even radios may be down or severely limited. Re-establishing communication is critical for connecting communities, sharing information, and coordinating efforts.

Radio networks: Survivors can set up local or regional radio networks using ham radios, walkie-talkies, or other low-tech communication devices. These networks allow communities to stay in touch, warn of dangers, and exchange vital information.

Messengers and couriers: In the absence of technology, communities may need to rely on messengers or couriers to deliver information and goods between nearby settlements. This primitive but effective method was used throughout history before modern communication systems.

Rebuilding Transportation

Transportation is essential for moving people, goods, and resources between communities. However, with roads damaged, vehicles destroyed, and fuel supplies limited, survivors will need to find alternative means of transport.

Bicycles and carts: Bicycles and carts powered by human or animal labor will likely become the primary means of transport, especially in areas where fuel is unavailable. These low-tech solutions are sustainable and can cover short to moderate distances.

Repairing roads: As communities expand, repairing and maintaining roadways will become a priority for trade and travel. This may involve clearing debris, filling potholes, and constructing simple bridges or crossings over rivers.

Renewable Energy and Power Sources

In the absence of large-scale power grids, communities will need to find alternative energy sources. Renewable energy solutions, such as solar, wind, and hydropower, can provide small-scale power for lighting, communication, and basic tools.

Solar power: Solar panels are relatively easy to maintain and can provide small amounts of electricity for basic needs like lighting, radios, and water pumps. Communities should salvage any available solar panels and batteries to create micro-power grids.

Wind and water power: If the community is located near a river or in a windy area, small wind turbines or water wheels can be constructed to generate electricity or mechanical power.

Trade and Bartering

In a post-nuclear world, traditional currency may lose its value, and the economy will likely shift to a barter system. Communities will trade goods and services directly, relying on the production of essential items like food, water, tools, and clothing.

Barter markets: Communities can set up local barter markets where individuals exchange goods and services. These markets allow for the distribution of resources based on need and availability.

Skills and labor: In addition to goods, skills and labor will become valuable commodities. People with specialized knowledge, such as blacksmiths, doctors, carpenters, or farmers, will be in high demand, and their services can be traded for food, shelter, or other necessities.

Re-establishing Governance and Law

As society begins to rebuild, questions of governance and law will inevitably arise. Communities will need to develop systems for making collective decisions, resolving disputes, and maintaining order.

Local Governance

In the early stages of rebuilding, governance will likely be local, based on small communities or clusters of survivors. Decisions about resource management, defense, and justice will be made at the community level, often through direct democracy or by trusted leaders.

Community councils: One possible model for governance is the formation of community councils, where representatives from different groups within the community gather to discuss and make decisions. These councils can be elected or appointed based on skills, trust, and leadership ability.

Shared decision-making: Some communities may adopt more collective decision-making processes, where all members have a voice in major decisions. This democratic approach ensures that everyone's needs and concerns are addressed.

Establishing Laws and Rules

As communities grow, they will need to establish laws or rules to maintain order and resolve conflicts. These laws may cover resource distribution, security, interpersonal disputes, and community responsibilities.

Basic laws: The first laws are likely to focus on protecting the community's survival. This includes laws about sharing resources, theft, violence, and maintaining hygiene and safety standards.

Enforcement: In the absence of a formal police force, law enforcement will likely be handled by community members themselves, possibly through a council or designated peacekeepers.

Dealing with Conflict

In any society, conflict is inevitable, and communities will need to develop methods for resolving disputes peacefully. This might involve mediation, community discussion, or formalized systems of justice.

Mediation: Mediation by trusted community members or leaders can help resolve interpersonal conflicts before they escalate into violence or resentment. Having a system in place for resolving disputes is essential for maintaining harmony.

Justice systems: Over time, communities may establish more formal justice systems to handle crimes or serious disputes. Punishments will likely focus on restorative justice, such as requiring individuals to contribute labor or resources to the community as a way of making amends.

Rebuilding Culture and Society

As communities stabilize and infrastructure is restored, attention will turn to rebuilding the social and cultural aspects of life. This includes education, arts, traditions, and a sense of identity and purpose.

Education and Skills Training

Education is key to rebuilding society. Survivors will need to pass on essential survival skills, such as farming, construction, and medicine, to future generations. As stability returns, more traditional forms of education can also resume, focusing on science, history, and technology.

Survival education: In the early stages, education will focus on practical survival skills. This includes teaching children and young adults how to grow food, purify water, and build shelters.

Skills training: Communities may set up training programs for essential skills, such as carpentry, blacksmithing, and medical care, ensuring that each person contributes to the community's survival.

Rebuilding Cultural Identity

Cultural traditions, arts, and social bonds will play a crucial role in rebuilding a sense of community and identity. Music, storytelling, and rituals can help survivors process trauma, remember their history, and create a sense of belonging in the new world.

Storytelling and oral history: In the absence of written records, storytelling and oral history will help preserve the community's past and pass down knowledge to future generations. Storytelling can also provide comfort and unity during difficult times.

Arts and crafts: As societies rebuild, arts and crafts will serve as both a creative outlet and a means of producing practical goods like clothing, tools, and decorations. Creative expression will also play an important role in healing from the trauma of nuclear disaster.

Religion and Spirituality

For many survivors, religion and spirituality will provide a sense of hope and purpose in a devastated world. Communities may turn to faith or spiritual practices to cope with loss, guide moral decisions, and rebuild social cohesion.

Community rituals: Shared religious or spiritual rituals can strengthen social bonds and provide a sense of normalcy in the face of chaos. These rituals may involve prayers for the dead, celebrations of life, or rituals of renewal as the community rebuilds.

Rebuilding society after a nuclear fallout is a long, arduous process that begins with survival and slowly evolves into the re-establishment of communities, infrastructure, governance, and culture. While the challenges will be immense, human resilience, cooperation, and ingenuity will play key roles in overcoming the obstacles. As survivors come together to form small, functioning communities, restore essential services, and create new systems of governance, they will lay the foundations for a future society. By focusing on cooperation, resource management, education, and cultural rebuilding, humanity can move beyond the fallout and toward a more hopeful future.

Radiation and Children: Protecting the Next Generation

In the aftermath of a nuclear disaster, children are among the most vulnerable to the harmful effects of radiation. Their developing bodies, smaller size, and higher rates of cell division make them more susceptible to radiation-related illnesses, long-term health effects, and environmental dangers. Protecting the next generation in a fallout world requires specific precautions, strategies, and careful planning to minimize their exposure to radiation and ensure they can grow and thrive in a hazardous environment. This chapter explores how radiation affects children differently from adults, the immediate and long-term health risks they face, and how to protect them physically, mentally, and emotionally.

Understanding the Impact of Radiation on Children

Radiation exposure affects children more severely than adults due to several physiological factors. Their growing bodies absorb radiation more readily, and the effects can disrupt their development, increase cancer risks, and cause genetic damage that may not become apparent until years later.

Why Children Are More Vulnerable

Several factors contribute to the heightened vulnerability of children to radiation exposure:

Faster cell division: Children's bodies are growing, and their cells divide more rapidly than those of adults. Radiation causes damage by disrupting cellular processes, and fast-dividing cells are more likely to become cancerous or develop abnormalities.

Smaller bodies: Children's smaller body mass means that radiation doses are more concentrated, increasing the overall impact on their organs and tissues.

Developing immune systems: Children's immune systems are still maturing, making it harder for them to fight off infections or illnesses caused by radiation exposure. This leaves them more vulnerable to both immediate radiation sickness and long-term health effects like cancer.

Absorption of contaminants: Children are more likely to absorb radioactive particles from the environment because they tend to engage in more physical play, may accidentally ingest contaminated soil, dust, or food, and often have closer contact with the ground.

Short-Term Effects of Radiation on Children

Radiation exposure can cause a range of immediate health effects in children, particularly if they are exposed to high levels in the immediate aftermath of a nuclear event.

Radiation sickness: Acute radiation syndrome (ARS) can occur when children are exposed to a high dose of radiation over a short period. Symptoms include nausea, vomiting, diarrhea, fatigue, and skin burns. Children are more prone to severe cases of ARS due to their smaller body size.

Burns and skin damage: Radiation can cause damage to the skin, leading to burns, lesions, or rashes. Children's thinner and more sensitive skin makes them particularly vulnerable to radiation burns.

Immediate immune suppression: Radiation weakens the immune system, which is particularly dangerous for children as it leaves them more susceptible to infections, illnesses, and slow recovery from injuries.

Long-Term Health Risks for Children

The long-term effects of radiation exposure are more severe for children due to their developmental stage. These effects may take years or even decades to manifest, but they significantly increase the risk of serious health problems.

Cancer: The most significant long-term risk of radiation exposure is an increased likelihood of developing cancers, especially leukemia and thyroid cancer. Since children have more time ahead of them for these cancers to develop, their lifetime cancer risk is significantly higher than that of adults exposed to the same radiation levels.

Genetic damage: Radiation can cause genetic mutations that may not affect the child directly but can lead to birth defects in future generations. These genetic changes can also increase the risk of chronic illnesses or developmental issues.

Growth and developmental delays: Radiation exposure, particularly at high levels, can disrupt normal growth and development, potentially leading to stunted growth, developmental delays, or cognitive impairments.

Protecting Children from Radiation Exposure

Given their heightened vulnerability, protecting children from radiation should be a top priority in a fallout environment. This involves both immediate actions to reduce exposure and long-term strategies to ensure their safety and health.

Creating Safe Shelters for Children

The first line of defense in protecting children from radiation is to create a safe, well-shielded shelter where they can remain during periods of heavy fallout. Ensuring that children spend as much time as possible indoors, away from direct radiation and contaminated surfaces, is crucial.

Indoor sheltering: Keep children indoors, particularly during the first few days or weeks after a nuclear event when radiation levels are highest. Ideally, your shelter should be underground or well-shielded by thick walls made of concrete, earth, or metal.

Designated play areas: Within your shelter, create designated safe zones where children can play, learn, and interact without risk of contamination. Ensure these areas are free from exposure to outside dust, dirt, or radioactive particles.

Air filtration: Use air filtration systems, such as HEPA filters, to keep indoor air as clean as possible. This will reduce the risk of inhaling radioactive particles that can accumulate indoors over time.

Clothing and Decontamination

Children will need protective clothing when they are outside or exposed to any potential fallout. Decontaminating them after exposure is also crucial to prevent the spread of radioactive particles into safe areas.

Protective clothing: Whenever children must leave the shelter, ensure they are dressed in protective gear such as long sleeves, pants, hats, and, if possible, radiation suits or masks. This clothing should be discarded or thoroughly cleaned after each use to avoid bringing contaminants inside.

Decontamination procedures: After spending time outside, children should be immediately decontaminated by removing all clothing and washing thoroughly with clean, uncontaminated water. Pay close attention to their hair, hands, and exposed skin. For small children, caregivers should assist to ensure thorough cleaning.

Ensuring Clean Food and Water for Children

Providing uncontaminated food and water is critical for keeping children healthy. Radiation can easily contaminate water sources and food, so precautions must be taken to ensure what they consume is safe.

Water purification: Use filtered, distilled, or bottled water for all drinking, cooking, and hygiene purposes for children. If you must use water from natural sources, always filter and purify it before use to remove radioactive particles.

Contaminant-free food: Avoid giving children food grown in contaminated soil or exposed to fallout. Instead, prioritize stored non-perishables, canned goods, and any food grown in controlled, indoor environments like greenhouses. If you are foraging or hunting, ensure the food is thoroughly cleaned and cooked to reduce contamination.

Nutritional supplements: If food supplies are limited, consider using multivitamins or supplements to ensure that children receive the essential nutrients needed for growth and development.

Radiation Monitoring and Health Checks

Regular monitoring of radiation levels and health checks are essential for protecting children from the harmful effects of radiation exposure. Early detection of exposure can help mitigate long-term health risks.

Radiation detection: Use radiation detectors, such as Geiger counters or dosimeters, to regularly test the environment for radiation levels. Monitor the areas where children spend most of their time and limit their exposure based on the readings. Ideally, radiation levels in safe zones should be below 0.1 μSv/h.

Health monitoring: Keep track of any changes in children's health, including signs of radiation sickness such as nausea, fatigue, skin rashes, or unusual bruising. Regularly check for weight loss, growth delays, or unusual behaviors that could indicate developmental issues.

Medical attention: If you have access to medical professionals, ensure that children receive regular health assessments to check for early signs of radiation exposure or related illnesses. If radiation sickness is suspected, seek medical care as soon as possible to prevent further complications.

Mental and Emotional Support for Children

In addition to protecting children physically, it's essential to support their mental and emotional well-being in the chaotic and frightening environment of a post-nuclear world. Children may struggle to understand what is happening, and the trauma of experiencing a disaster can have lasting psychological effects.

Providing Comfort and Stability

Children thrive on routine and stability, even in extreme circumstances. Establishing a daily routine can help them feel more secure and provide a sense of normalcy amidst the chaos.

Daily routines: Create a structured daily routine that includes time for learning, play, and rest. Even in a survival scenario, maintaining some consistency can help reduce anxiety and provide a sense of control for children.

Emotional reassurance: Children will look to adults for reassurance during uncertain times. Be honest with them about the situation but offer comfort and reassurance whenever possible. Encourage open conversations where children can express their fears and feelings.

Creating a Safe Emotional Environment

In a stressful post-nuclear environment, children may experience heightened anxiety, fear, or depression. Helping them process these emotions in a healthy way is essential for their long-term mental health.

Creative outlets: Provide opportunities for children to express their emotions through creative outlets like drawing, writing, or storytelling. This allows them to process their experiences in a constructive manner and can be therapeutic.

Physical play: Encourage physical play and movement, even in small indoor spaces. This helps release stress and pent-up energy, which is particularly important for younger children who may struggle to sit still for long periods.

Positive social interactions: Foster positive social interactions between children and adults or other children in the shelter. This helps build a sense of community and reduces feelings of isolation or loneliness.

Long-Term Psychological Support

Children who survive a nuclear disaster will likely experience long-term psychological impacts, such as post-traumatic stress disorder (PTSD) or anxiety about the future. Providing ongoing emotional support is crucial to their recovery.

Counseling and therapy: If possible, provide access to psychological counseling or therapy, especially for children who show signs of trauma. In the absence of professional mental health support, create a safe space for children to talk about their fears, experiences, and hopes for the future.

Education about the future: Helping children understand that life can continue, even after a disaster, can give them hope. Teach them survival skills and emphasize the importance of resilience. As the environment stabilizes, focus on building a sense of purpose for the future.

Preparing Children for a Post-Nuclear Future

Ultimately, protecting children in a post-nuclear world means preparing them to survive and thrive in the new reality. Teaching them practical survival skills, fostering resilience, and giving them the tools to navigate the dangers of a fallout environment are crucial steps in ensuring the next generation's survival.

Teaching Survival Skills

As children grow, they will need to learn how to take care of themselves and contribute to the community. Teaching age-appropriate survival skills can empower them and give them a sense of purpose.

Basic survival skills: Teach children how to find and purify water, grow food, forage safely, and defend themselves. These skills will become essential as they mature and take on more responsibility in a survival community.

Radiation safety: Help children understand the dangers of radiation and how to avoid exposure. Teach them how to use protective gear, follow decontamination procedures, and recognize contaminated areas.

Building Resilience

In the face of adversity, resilience is one of the most valuable qualities children can develop. Help them build mental and emotional resilience by encouraging problem-solving, teamwork, and perseverance.

Problem-solving: Encourage children to take part in solving daily challenges, whether it's helping with food preparation, fixing equipment, or organizing supplies. This helps build confidence and fosters a sense of independence.

Teamwork: Emphasize the importance of working together as a community. Children should learn to rely on others while also contributing to the group's well-being.

Planning for the Future

Despite the immediate challenges of living in a fallout environment, it's important to instill hope for the future in children. Teach them that they are part of rebuilding society and that their efforts can create a better world.

Education and knowledge: As society stabilizes, ensure that children receive an education, even if it's informal. Learning to read, write, and understand science will prepare them for rebuilding society and advancing technology.

Hope and purpose: Encourage children to think about the future, even in difficult circumstances. Talk about rebuilding homes, schools, and communities once the environment is safer, and give them tasks that contribute to the long-term survival and success of the group.

Protecting children from the dangers of radiation and the hardships of a fallout world is a critical aspect of long-term survival after a nuclear disaster. The challenges are immense, but with careful planning and a focus on both short-term safety and long-term development, children can survive and even thrive in a post-nuclear world, growing into the leaders and builders of a renewed society.

First Responders in Nuclear Crises: What to Expect

In the immediate aftermath of a nuclear disaster, the role of first responders is both critical and fraught with immense challenges. These individuals—emergency medical teams, firefighters, law enforcement, military personnel, and volunteers—are tasked with saving lives, stabilizing dangerous situations, and coordinating the initial response to the fallout. However, nuclear crises are unique in their scale, complexity, and the extreme hazards they present, including radiation exposure, mass casualties, and infrastructure collapse. Understanding what to expect from first responders in a nuclear crisis can help you better prepare for their arrival (if they come) and know how to collaborate or act independently if the situation requires it.

This chapter will examine the expected role of first responders during nuclear disasters, their priorities, the limitations they face, and what individuals and communities should be prepared for if help arrives or if they must take action on their own.

The Role of First Responders in a Nuclear Crisis

In any disaster, first responders are responsible for providing immediate assistance to those affected, stabilizing hazardous environments, and coordinating the initial stages of recovery. In a nuclear crisis, their role expands to include managing radiation exposure, dealing with mass casualties, and attempting to restore basic services in an environment that is likely overwhelmed by chaos and destruction.

Emergency Medical Response

Emergency medical responders will be among the first on the scene, focusing on saving lives, treating injuries, and managing the health impact of radiation exposure.

Triage: In a nuclear crisis, first responders must prioritize patients based on the severity of their injuries and the likelihood of survival. Triage systems will be used to categorize victims into groups such as those requiring immediate attention, those who can wait, and those whose injuries are too severe to survive with the available resources.

Treating radiation sickness: Radiation exposure causes acute radiation syndrome (ARS) in severe cases. First responders will treat symptoms such as nausea, vomiting, and burns, but specialized medical care will be needed for serious exposure cases. The ability to provide this care may be limited, especially if medical infrastructure has been damaged.

Transporting the injured: In heavily affected areas, medical evacuation may be necessary. First responders will coordinate the transport of critically injured individuals to hospitals or field medical stations, though overwhelmed infrastructure may slow this process.

Firefighting and Hazardous Materials (HAZMAT) Response

In addition to the damage caused by the nuclear blast itself, fires, chemical spills, and secondary explosions may occur. Firefighters and hazardous materials teams will work to contain these risks, prevent further spread of fallout, and protect survivors from additional harm.

Extinguishing fires: Fires are common in the wake of nuclear explosions due to the intense heat and shockwave. Firefighters will attempt to extinguish fires, especially near key infrastructure like hospitals, power plants, and evacuation centers, to prevent further casualties.

HAZMAT teams: Specialized hazardous materials teams will focus on containing and decontaminating areas where dangerous substances, including radioactive fallout, have been released. These teams will also attempt to secure or seal off critical infrastructure, such as nuclear power plants, to prevent additional leaks or explosions.

Controlling the spread of fallout: Part of the HAZMAT response will include monitoring radiation levels and attempting to contain fallout contamination in urban or populated areas. This may involve decontaminating key areas like hospitals, schools, or shelters to ensure they are safe for survivors to occupy.

Law Enforcement and Crowd Control

Law enforcement plays a vital role in maintaining order, ensuring public safety, and coordinating evacuation efforts. In a nuclear disaster, civil unrest, panic, and looting are likely, and law enforcement will need to manage these situations while prioritizing public safety.

Maintaining order: In the chaos following a nuclear disaster, panic, looting, and violence can occur as survivors scramble for resources. Law enforcement's primary role will be to maintain order, prevent crime, and protect both civilians and critical infrastructure.

Evacuation coordination: Police and military forces will likely be involved in managing evacuations, directing survivors to safe zones, and coordinating transportation out of high-risk areas. They may establish checkpoints to control movement and prevent individuals from entering dangerous zones.

Protecting critical infrastructure: Law enforcement will also be responsible for securing critical infrastructure such as hospitals, communication centers, and transportation hubs. These sites will become focal points for relief efforts and will need protection from both physical threats and looting.

Search and Rescue Operations

Search and rescue teams will play a critical role in locating and saving survivors trapped under debris or in areas affected by radiation. These teams often work alongside emergency medical personnel to extract individuals from dangerous situations and provide immediate medical care.

Locating survivors: In the aftermath of a nuclear explosion, buildings may be damaged or collapsed, trapping individuals under debris. Search and rescue teams will focus on locating these survivors using tools like thermal imaging, drones, and search dogs.

Dealing with radiation zones: Many search and rescue teams will be equipped with radiation detectors to avoid entering highly radioactive areas or to work safely within them for short periods. Specialized training allows them to extract survivors while minimizing their own radiation exposure.

Medical assistance: Search and rescue teams will often provide immediate first aid to survivors once they are located, particularly if medical teams are delayed in reaching the scene.

Military Involvement and Martial Law

In the event of a large-scale nuclear disaster, military forces may be mobilized to assist with rescue operations, distribute aid, and maintain order. In extreme situations, martial law may be declared, giving military forces direct control over civil functions.

Evacuations and security: The military may establish and enforce evacuation zones, protect key infrastructure, and provide security at shelters or relief centers. They will also assist law enforcement in maintaining order and controlling access to high-risk areas.

Aid distribution: Military personnel will likely play a key role in distributing food, water, medical supplies, and other resources to survivors. They may also set up temporary shelters and field hospitals in safe zones to provide emergency care.

Martial law: If civil authorities are overwhelmed or unable to maintain control, the government may declare martial law, placing the military in charge of maintaining order, enforcing curfews, and controlling access to resources. This can include restricting civilian movement and temporarily suspending certain civil liberties to ensure public safety.

Challenges and Limitations Facing First Responders

First responders in a nuclear crisis face immense challenges that can limit their ability to assist survivors effectively. The scale of destruction, radiation exposure, damaged infrastructure, and the sheer number of casualties can overwhelm even the best-prepared response teams. Survivors should understand these limitations and be prepared to take independent action if help is delayed or unavailable.

Radiation Exposure

One of the most significant challenges for first responders is the danger of radiation exposure. While they may be equipped with protective gear, radiation levels in the immediate aftermath of a nuclear event can be extremely high, limiting how long they can safely operate in affected areas.

Protective gear: First responders will likely be wearing radiation suits and using dosimeters to monitor their exposure. However, even with protection, their ability to operate in high-radiation zones will be limited by the need to minimize their own exposure.

Evacuation of high-radiation zones: In some cases, first responders may be unable to enter certain areas until radiation levels drop, which could delay rescue efforts. Survivors in these zones may need to evacuate themselves or wait for assistance.

Infrastructure Damage

Nuclear explosions cause widespread infrastructure damage, including the destruction of roads, bridges, communication networks, and hospitals. First responders will need to navigate these challenges to reach survivors and coordinate their efforts.

Transportation challenges: Blocked roads, collapsed bridges, and damaged transportation networks will slow down the arrival of aid and evacuation efforts. Survivors may be cut off from help for extended periods.

Communication breakdown: Damaged communication systems will make it difficult for first responders to coordinate their efforts or relay important information to the public. Survivors may not receive timely updates on evacuation plans or safe zones.

Overwhelming Casualties

The sheer number of casualties in a nuclear crisis can overwhelm medical teams, hospitals, and rescue operations. First responders will need to make difficult decisions about how to allocate limited resources.

Triage limitations: In a mass casualty situation, first responders may be forced to focus on those most likely to survive, leaving severely injured individuals without treatment. Survivors should be prepared to take basic first aid measures on their own if medical help is delayed.

Field hospitals and makeshift care: If hospitals are overwhelmed or destroyed, first responders may set up field hospitals to provide basic care. However, these makeshift facilities will likely have limited resources and may prioritize stabilization over long-term treatment.

Limited Resources and Supplies

First responders will face significant resource shortages in a nuclear crisis. Medical supplies, protective gear, food, and water will be in short supply, and distribution may be delayed due to infrastructure damage and communication breakdowns.

Medical shortages: Hospitals and medical teams may run out of critical supplies like antibiotics, radiation treatments, bandages, and IV fluids. This may limit their ability to treat radiation sickness, burns, or traumatic injuries.

Food and water distribution: First responders will prioritize distributing food and water to survivors, but shortages may occur. Survivors should be prepared to ration their own supplies until aid reaches them.

What to Expect from First Responders

While first responders will do everything in their power to assist survivors, nuclear crises are unpredictable and challenging. It's important to know what to expect and how to prepare for potential delays or limitations in the response.

Prioritization of Critical Areas

First responders will likely prioritize areas with the highest concentration of survivors or critical infrastructure, such as hospitals, schools, and major evacuation zones. If you are in a more remote area, it may take longer for help to arrive.

Delayed Response Times

Radiation levels, blocked roads, and communication breakdowns will likely slow down response times. Survivors should be prepared to shelter in place or evacuate on their own if help is delayed.

Focus on Stabilization and Evacuation

In the early stages of the crisis, first responders will focus on stabilizing the situation—extinguishing fires, treating injuries, and evacuating survivors from dangerous areas. Long-term recovery efforts may not begin until the immediate crisis is under control.

Preparing to Act Independently

Given the limitations and challenges that first responders face in a nuclear crisis, it's important for individuals and communities to be prepared to act independently if necessary. This includes having the skills, resources, and knowledge to survive without immediate help.

First Aid and Medical Supplies

Ensure you have basic first aid supplies, including bandages, antiseptics, pain relievers, and, if possible, potassium iodide tablets to protect against radiation exposure. Learn basic first aid techniques to stabilize injuries until help arrives.

Radiation Protection

Prepare by having protective gear, such as dust masks, long sleeves, pants, and, if available, radiation suits. Understand how to decontaminate yourself and your environment after exposure to radioactive fallout.

Self-Evacuation Plans

If you are in a high-risk area, be prepared to evacuate on your own. Know the safest routes away from the fallout zone, and have a plan for where to go, whether to a government-designated shelter or a prearranged safe location.

Long-Term Survival Supplies

Stockpile food, water, and essential supplies for at least several weeks. In a nuclear disaster, aid may be delayed, and it could be a long time before normal supply chains are restored.

First responders will play a vital role in managing the immediate aftermath of a nuclear crisis, but their ability to help may be limited by the unique challenges posed by radiation, infrastructure damage, and overwhelming casualties. While they will focus on saving lives, stabilizing dangerous situations, and providing aid, survivors must be prepared to take action independently if help is delayed or unavailable. Understanding the role of first responders, their priorities, and the challenges they face can help you better prepare for the realities of surviving and rebuilding in the aftermath of a nuclear disaster.

Dealing with Fallout on a Budget: Survival Tips for the Frugal Prepper

In the event of a nuclear disaster, preparing for survival doesn't have to break the bank. For many, prepping on a budget is a necessity, but with thoughtful planning, resourcefulness, and an understanding of key survival principles, even the frugal prepper can effectively prepare to deal with fallout. This chapter focuses on practical, cost-effective tips and strategies to prepare for a nuclear fallout while staying within a tight budget. From building affordable shelters to creating low-cost water filtration systems and stockpiling inexpensive food, we'll explore ways to stretch your resources and maximize your chances of survival.

The Importance of Budget-Friendly Preparedness

Preparedness doesn't have to be expensive. The key to surviving nuclear fallout lies in making smart choices with your limited resources. By focusing on affordable solutions and DIY strategies, you can build a robust survival plan that protects you and your family without requiring high-cost investments. Frugal prepping is about understanding the essentials and prioritizing the most critical needs—shelter, clean water, food, and protection from radiation.

Prioritizing Essentials on a Budget

The key to survival in a fallout scenario is addressing the fundamental needs: shelter, water, food, and safety from radiation. Here's how to tackle each of these priorities without spending a fortune.

Affordable Fallout Shelter Solutions

Having a safe space to shelter during the initial fallout is critical, and while building a high-end bunker is ideal, it's often not financially feasible for the average prepper. However, there are budget-friendly ways to protect yourself from radiation and the elements.

Use existing spaces: If you don't have the funds to build an underground bunker, consider using parts of your home, such as basements, interior rooms, or even closets. Basements offer excellent protection because they are below ground level, reducing exposure to radiation. Reinforce basement walls with sandbags or concrete blocks if possible to increase shielding.

DIY radiation shielding: To create a more affordable shelter, add layers of thick materials like dirt, sandbags, concrete blocks, or even stacked books to absorb radiation. The more mass between you and the fallout, the better protected you'll be. Thick walls or makeshift barriers can help significantly reduce exposure.

Build an indoor safe room: Even if you lack a basement, you can build a small "safe room" within your home. Use heavy furniture or bookshelves to reinforce walls, and fill them with dense materials to act as barriers. Line windows and doors with plastic sheets, duct tape, or tarps to keep radioactive particles out.

Create a "lean-to" bunker: If you have outdoor space and minimal funds, consider constructing a lean-to shelter. Dig a trench that's about 2-3 feet deep, and line it with dirt or sandbags for radiation shielding. Cover the top with sturdy material like wooden planks or metal sheeting and pile on dirt for additional protection.

Water Purification on a Budget

Clean water is essential in any disaster scenario, and while purchasing expensive water filtration systems is ideal, there are affordable alternatives to ensure safe drinking water.

Boiling water: One of the simplest and most cost-effective ways to purify water is by boiling it. If you can boil water, it will kill most bacteria and pathogens. While boiling won't remove radioactive particles, it will make water safer to drink when filtered first.

DIY filtration: You can create a homemade water filtration system using common materials. A simple filter can be made by layering sand, gravel, and charcoal in a plastic bottle or bucket with holes at the bottom. This DIY filter will help remove particles and sediment, though it's still essential to purify the water afterward, such as through boiling or chemical treatment.

Bleach disinfection: Unscented household bleach can be used to disinfect water. Add 8 drops of bleach per gallon of water, stir, and let it sit for 30 minutes before drinking. Make sure the bleach contains only sodium hypochlorite (at 6%) and no additional chemicals like fragrance or dyes.

Rainwater collection: Collecting rainwater is another low-cost solution. Set up tarps, clean plastic sheeting, or gutters to direct rainwater into clean containers. Always purify the collected water before drinking.

Water storage on a budget: Repurpose large plastic containers, such as soda bottles or food-grade buckets, to store clean water. Make sure they are properly sanitized before use, and label the containers to track when they were filled.

Low-Cost Food Stockpiling

Food storage is essential for long-term survival, but it doesn't need to be expensive. Focus on buying cheap, non-perishable items that provide a good balance of calories and nutrition.

Inexpensive staples: Some of the most cost-effective foods for survival include rice, dried beans, lentils, oats, pasta, and flour. These staples are calorie-dense, have long shelf lives, and can be bought in bulk at low cost.

Canned goods: Stockpile affordable canned goods, including vegetables, beans, meat, and fruit. Canned food can last for years and is an excellent backup for fresh or frozen food. Look for sales or buy in bulk to save money.

DIY preservation: Learn basic food preservation techniques, such as dehydrating, smoking, or fermenting, to extend the life of fresh produce and meat. A solar dehydrator or homemade smoker can be made using cheap or salvaged materials, and these methods allow you to preserve food without electricity.

Foraging and gardening: Growing your own food is an affordable way to supplement your stockpile. Even in a fallout environment, small indoor gardens or container gardening can produce leafy greens, herbs, and root vegetables. Foraging for wild edible plants and berries is another cost-free source of nutrition, but ensure the environment is safe from radiation contamination before consuming foraged food.

Rationing for long-term survival: Plan to ration your food carefully to make it last. Reduce portion sizes, combine low-cost staples with high-nutrition items like canned protein or dried fruits, and focus on nutrient-dense meals that keep you full longer.

Radiation Protection on a Budget

Radiation protection is a priority, but full radiation suits and professional-grade protective gear are expensive. Fortunately, there are more budget-friendly ways to protect yourself from fallout.

DIY protective gear: You can improvise protective clothing using items you already own. Wear long-sleeved shirts, pants, hats, and gloves to cover exposed skin. Use heavy-duty plastic sheeting or garbage bags over clothing to create a protective barrier, and wear goggles or a dust mask to prevent inhaling radioactive particles.

Decontamination methods: If you're exposed to fallout, decontaminate yourself immediately by removing clothing and washing your skin thoroughly with soap and clean water. Keep a supply of wet wipes or moist towelettes on hand for quick decontamination if water is scarce.

Indoor air quality: Use plastic sheeting, tarps, or garbage bags to seal windows, doors, and vents to keep radioactive particles from entering your living space. If you don't have access to expensive air filters, make DIY air purifiers using box fans with HEPA filters taped to the back.

Potassium iodide (KI) tablets: If you can't afford large quantities of expensive radiation treatments, stock up on affordable potassium iodide tablets, which protect the thyroid gland from radioactive iodine in the event of fallout. These tablets are low-cost and easy to store.

Self-Sufficiency and DIY Solutions

In a post-fallout world, self-sufficiency is crucial. Learning practical, budget-friendly skills and DIY solutions can help you survive without expensive tools or supplies.

Bartering for supplies: Build a network of like-minded preppers or neighbors with whom you can barter goods and services. If you have skills in carpentry, farming, or mechanics, you can trade labor for food, medical supplies, or other necessities.

Repurpose and salvage materials: Scavenge and repurpose materials like wood, metal, and plastic to create tools, repair structures, or build storage solutions. Old furniture, scrap metal, and other discarded items can often be used to reinforce shelters, make containers, or create simple devices like solar cookers.

DIY tools: Learn to craft basic tools from available materials. For example, a solar oven can be made from cardboard, foil, and glass or plastic, allowing you to cook food without fuel. Similarly, you can create improvised fishing gear or hunting traps using everyday materials like string, wire, and wood.

Budget lighting solutions: Use solar-powered garden lights or crank-powered flashlights as cost-effective lighting options. They are inexpensive, long-lasting, and don't require batteries or electricity, making them ideal for survival situations.

Health and Hygiene on a Budget

Maintaining health and hygiene is critical in a fallout scenario, and you can prepare for this without spending a lot.

Inexpensive first aid supplies: Stock up on basic first aid supplies like bandages, antiseptics, pain relievers, and gauze. These can often be found at discount stores or bought in bulk for a low cost. Focus on items that can address common injuries, cuts, and burns.

DIY hygiene products: You can make your own soap using cheap ingredients like lye and fat, or stockpile inexpensive bar soap, which has a long shelf life. Baking soda can be used as toothpaste, a deodorant, and a cleaning agent.

Improvised sanitation: Set up basic sanitation facilities using plastic buckets, garbage bags, and sawdust for an improvised toilet. This keeps waste contained and prevents contamination of your living area. Have a supply of cheap bleach or disinfectants to clean surfaces and control germs.

Long-Term Survival Strategies for the Frugal Prepper

Surviving nuclear fallout is about more than just immediate shelter and supplies—it's about long-term sustainability. Here are some budget-friendly strategies to prepare for the long haul:

Learning Survival Skills

Invest time in learning essential survival skills such as food preservation, first aid, foraging, and water purification. Knowledge is often the most valuable resource in a survival situation, and learning these skills costs little or nothing.

Building a Community

Establish a network of like-minded preppers or neighbors who can share resources, knowledge, and labor. Community-building allows you to pool resources, barter, and offer mutual support in times of need.

Self-Reliance Through Homesteading

Homesteading—growing your own food, raising livestock, and creating renewable resources—is one of the most sustainable ways to survive on a budget. Even if you have limited space or funds, starting a small garden, keeping chickens, or learning to compost can significantly enhance your long-term survival prospects.

Bartering and Trade

Consider learning a valuable trade skill such as carpentry, blacksmithing, or basic mechanics. These skills can be used for bartering in a post-fallout world when traditional currency may lose its value.

Surviving nuclear fallout on a budget requires resourcefulness, careful planning, and the willingness to use what's available. With the right strategies, you can protect yourself from radiation, secure food and water, and create a sustainable survival plan without spending a fortune. By focusing on affordable solutions, DIY methods, and building a self-reliant lifestyle, even the most frugal prepper can be well-prepared for the challenges of a post-nuclear world.

Learning from History: Nuclear Disasters and Survival Lessons

Throughout history, humanity has witnessed several nuclear disasters, each offering critical survival lessons for the future. From the bombings of Hiroshima and Nagasaki to nuclear plant meltdowns like Chernobyl and Fukushima, these events have shaped our understanding of radiation, fallout, and the immense challenges faced by survivors. Studying these past nuclear crises can provide valuable insights into what works—and what doesn't—in terms of preparation, survival strategies, and long-term recovery.

This chapter delves into the key nuclear disasters of history, exploring what we've learned from them, and how their lessons can be applied to surviving a future nuclear fallout. By analyzing these events, preppers can gain practical knowledge on how to deal with radiation, manage fallout, and navigate the physical, emotional, and societal aftermath of a nuclear catastrophe.

The Bombings of Hiroshima and Nagasaki (1945)

The atomic bombings of Hiroshima and Nagasaki during World War II marked the first—and only—use of nuclear weapons in warfare. These two catastrophic events killed hundreds of thousands of people and left lasting impacts on survivors, known as "hibakusha," who faced radiation sickness, cancer, and societal challenges for years after the bombings.

Survival Lessons from Hiroshima and Nagasaki

While Hiroshima and Nagasaki are unique due to their wartime context, they provide several key lessons on surviving a nuclear blast and its immediate aftermath:

Distance from the blast: The farther one was from the blast center, the greater the chances of survival. Those who were able to take shelter in sturdy buildings or underground structures had significantly higher survival rates. This underscores the importance of seeking immediate shelter if a nuclear event occurs.

Immediate sheltering: Many survivors of the bombings took cover in basements, concrete buildings, or behind thick barriers. These structures provided protection not only from the initial blast but also from the heat and radiation that followed. The key lesson is that the first few minutes after a blast are critical—finding the nearest, sturdiest shelter can make the difference between life and death.

Decontamination: After the bombs detonated, survivors who were able to wash off fallout particles (such as soot, ash, and radioactive dust) quickly reduced their exposure to radiation. A key takeaway is to decontaminate as soon as possible after fallout exposure to minimize radiation absorption.

Long-term health risks: Survivors of Hiroshima and Nagasaki faced long-term health consequences, particularly cancer and radiation-related illnesses. Regular medical checkups and monitoring for symptoms of radiation exposure are essential for anyone exposed to nuclear fallout.

The Chernobyl Disaster (1986)

The Chernobyl nuclear disaster in Ukraine remains one of the worst nuclear accidents in history. A reactor at the Chernobyl Nuclear Power Plant exploded, releasing massive amounts of radioactive materials into the environment. The long-term consequences of the disaster, both environmental and human, have provided important lessons for nuclear safety, evacuation, and radiation exposure management.

Survival Lessons from Chernobyl

Chernobyl taught us how a nuclear disaster at a power plant can have far-reaching effects, with radiation spreading across borders and affecting millions. Key lessons from Chernobyl include:

Evacuation and timing: The town of Pripyat, located near Chernobyl, was not evacuated until nearly two days after the explosion. Many residents were exposed to dangerous levels of radiation before authorities issued the evacuation order. The lesson here is that early evacuation is critical. If you are in a radiation zone and evacuation orders are not immediate, make the decision to leave on your own to reduce exposure.

Sheltering in place: While evacuation is ideal, there are times when it may not be possible. In Chernobyl, residents who stayed indoors during the initial fallout and sealed their windows and doors were better protected from radiation exposure than those who remained outside. The takeaway is that if evacuation is delayed or impossible, sheltering in place with protective measures can significantly reduce exposure.

Radiation monitoring and cleanup: After the explosion, massive efforts were made to monitor radiation levels and clean up contaminated areas. Although individuals cannot replicate large-scale government efforts, preppers can learn from this by ensuring they have access to radiation detectors and protective gear. Regular monitoring of radiation levels can help avoid highly contaminated areas.

Food and water contamination: Radiation spread across agricultural areas, contaminating food and water supplies. Chernobyl survivors had to be careful about what they consumed. The lesson here is to have a stockpile of uncontaminated food and water, and to use filtration or purification methods to ensure that drinking water is safe.

Long-term effects: The long-term health effects of the Chernobyl disaster include an increase in thyroid cancer and other radiation-induced diseases, particularly in children. Monitoring health and being vigilant about long-term consequences are critical components of survival after radiation exposure.

The Fukushima Daiichi Nuclear Disaster (2011)

In 2011, a massive earthquake and tsunami triggered the meltdown of reactors at the Fukushima Daiichi Nuclear Power Plant in Japan. The event caused widespread radiation contamination, forced large-scale evacuations, and resulted in long-lasting environmental and health impacts. Fukushima highlighted the risks associated with natural disasters and nuclear power, and it underscored the importance of preparedness and quick action in mitigating damage.

Survival Lessons from Fukushima

The Fukushima disaster provides key insights into how to handle nuclear emergencies compounded by natural disasters:

Preparedness for natural disasters: Fukushima teaches the importance of preparing for multiple, compounding disasters. Nuclear accidents may be triggered by natural events like earthquakes or floods, making it essential to have a comprehensive survival plan that accounts for a variety of scenarios.

Evacuation planning: Japanese authorities ordered the evacuation of people living within a 20-kilometer radius of the Fukushima plant. Many survivors faced chaos and confusion during the evacuation. The lesson here is that having a personal evacuation plan, knowing escape routes, and keeping an emergency kit ready can help avoid panic and ensure a smoother evacuation process.

Radiation spread by wind and water: One of the challenges with Fukushima was the spread of radioactive materials by both wind and water. This highlights the importance of understanding how fallout spreads and taking protective measures accordingly. Living downwind or near water sources connected to a contaminated area increases the risk of radiation exposure.

Dealing with contaminated food and water: After Fukushima, radiation spread to agricultural regions and fishing areas, contaminating crops and seafood. Survivors had to rely on alternative food sources. This reinforces the importance of having a reliable, uncontaminated food and water stockpile, particularly in regions near nuclear power plants or potential targets of nuclear attacks.

Communication breakdowns: During the Fukushima crisis, there were challenges in communication between government agencies, power plant operators, and the public. In a nuclear disaster, communication breakdowns are common, so having a plan for staying informed—such as a battery-operated radio or local networks—can help you receive critical information.

The Three Mile Island Incident (1979)

The Three Mile Island incident, which took place in Pennsylvania, USA, is often considered the most serious nuclear accident in the United States. Although it did not result in the widespread release of radiation like Chernobyl or Fukushima, the partial meltdown of a reactor led to widespread fear, evacuation orders, and debates about nuclear safety.

Survival Lessons from Three Mile Island

Though the Three Mile Island disaster was contained before it escalated, it offers important survival lessons, particularly regarding public perception, preparedness, and emergency response:

Public preparedness and panic: Even though the radiation release from Three Mile Island was minimal, confusion and fear caused panic in the surrounding population. Thousands of people evacuated despite no official order. This incident underscores the importance of having clear, reliable information in a nuclear crisis. Avoid panic by staying informed, verifying facts, and preparing for both evacuation and sheltering in place, depending on the situation.

Trust but verify government information: The government initially downplayed the seriousness of the incident, which led to mistrust among the public. Preppers should be ready to trust official information, but also take independent action if they believe the situation warrants it. Having access to independent tools, like a Geiger counter, can help you verify radiation levels for yourself.

Early warning systems: Although Three Mile Island didn't release high levels of radiation, the scare emphasized the importance of early warning systems. Today, many governments and communities near nuclear plants have implemented emergency alert systems. Knowing how to receive alerts and where to find information in an emergency is crucial for survival.

Lessons from the Cold War Era

During the Cold War, the threat of nuclear war led to widespread public preparation for potential fallout, especially in the United States and Soviet Union. Governments constructed fallout shelters, educated the public on nuclear safety, and developed plans for surviving a nuclear war. While we didn't experience global nuclear devastation during the Cold War, the period provides valuable lessons for today's preppers.

Survival Lessons from the Cold War

The Cold War era taught us that preparation and education are key to surviving a nuclear event:

Fallout shelters: Many people built personal fallout shelters during the Cold War, either at home or in public spaces. These shelters were designed to protect against radiation and blast effects. While modern preppers may not need Cold War-style bunkers, the principle of having a safe, well-stocked shelter remains vital.

Education and drills: Schools and communities conducted nuclear drills and provided citizens with information on how to survive a nuclear attack. Preppers today can learn from these efforts by educating themselves, conducting family drills, and having detailed evacuation and shelter plans.

Stockpiling necessities: The Cold War era emphasized stockpiling essentials like food, water, and medical supplies. Stockpiling remains one of the most effective ways to prepare for a nuclear disaster, ensuring that you can remain self-sufficient for an extended period.

Long-term preparedness: Cold War preparations were not short-term; people understood that recovery from a nuclear war could take years or even decades. Today's preppers should focus not only on immediate survival but also on long-term sustainability, including growing food, generating energy, and rebuilding communities.

Nuclear disasters, both man-made and natural, have taught humanity valuable lessons about survival, preparedness, and resilience. Whether it's the immediate destruction of a nuclear blast or the long-term consequences of radiation exposure, studying these historical events helps us understand what works and what doesn't in surviving such catastrophes.

Firearms and Defense in a Post-Apocalyptic World

In a post-apocalyptic world, where the breakdown of law and order is a real possibility, self-defense becomes one of the most critical components of survival. While there are many ways to defend yourself and your community, firearms are often considered the most effective tools for protecting against both human and animal threats. In a world where resources are scarce, tensions run high, and violence may become commonplace, having the right firearms—and knowing how to use them safely and responsibly—can make a significant difference in your chances of survival.

This chapter explores the role of firearms in a post-apocalyptic world, covering the types of firearms best suited for survival, practical considerations for using and maintaining them, and the ethical and legal aspects of using firearms for defense. Additionally, it addresses how to create a comprehensive defense strategy that includes more than just weaponry, emphasizing the importance of community, fortification, and preparedness.

The Role of Firearms in Post-Apocalyptic Survival

In a post-apocalyptic scenario, firearms serve several critical purposes, including protection from hostile individuals, hunting for food, and defense against wild animals. They can also serve as tools for intimidation, helping to deter threats without the need for direct confrontation. However, firearms are just one part of a larger survival strategy, and they come with responsibilities and potential risks that should be carefully considered.

Self-Defense Against Human Threats

In the chaos following a nuclear disaster or other apocalyptic event, civil order may break down, leading to increased risks of looting, banditry, and violent conflicts over resources. Firearms provide a means of protecting yourself, your family, and your community from those who may try to take what you have by force.

Deterrence: The mere presence of a firearm can often deter potential aggressors. Many individuals will think twice before attacking someone they know is armed, especially if they are unarmed themselves.

Close and long-range defense: Firearms allow you to defend yourself from both close-quarters attacks and long-range threats. Whether dealing with intruders trying to enter your home or encountering hostile groups in the open, the ability to engage from a distance provides a significant tactical advantage.

Protection from Wild Animals

In a post-apocalyptic environment, you may face threats from wild animals, especially if you live in rural or wilderness areas. As human populations diminish or retreat, animals may roam more freely into formerly populated regions, and hunting or scavenging for food could lead to dangerous encounters with predators.

Predator defense: Larger predators such as bears, wolves, or feral dogs may pose a significant threat if food becomes scarce. Firearms offer a practical means of defending yourself from animal attacks when avoidance or deterrence is not an option.

Livestock protection: If you are raising livestock for food, firearms can help protect your animals from predators that may prey on them, ensuring your food supply remains intact.

Hunting for Food

In a world where grocery stores are no longer an option, hunting may become one of your primary methods of acquiring food. Firearms, particularly rifles, are highly effective tools for hunting game, from small animals like rabbits and squirrels to larger animals like deer or elk.

Sustainable hunting: Knowing how to hunt efficiently and responsibly will allow you to provide for your family without exhausting local wildlife populations. The ability to hit targets at a distance with precision can reduce the need for long pursuits and conserve energy.

Ammunition management: In a post-apocalyptic world, ammunition may become a finite resource, so it's essential to hunt wisely. Take only the shots you are confident will be successful, and consider alternative methods of hunting (such as trapping or bowhunting) for smaller game to conserve ammunition.

Choosing the Right Firearms for Survival

Not all firearms are equally suited to a post-apocalyptic environment. The best firearms for survival are those that are versatile, reliable, and easy to maintain. Below are some of the key types of firearms to consider, along with their advantages and disadvantages.

Rifles

Rifles are the most versatile firearms for survival, offering range, power, and precision. They are ideal for hunting, defending a homestead, and engaging threats at a distance.

Bolt-action rifles: Bolt-action rifles are known for their reliability and accuracy, making them excellent choices for hunting and long-range defense. They are simple to maintain and often have fewer parts that can malfunction compared to semi-automatic rifles.

Semi-automatic rifles: Semi-automatic rifles offer a higher rate of fire and quicker follow-up shots than bolt-action rifles. Popular models like the AR-15 or AK-47 provide versatility for both defense and hunting. However, semi-automatic rifles require more maintenance and can be more complex to repair.

Lever-action rifles: Lever-action rifles are a good middle ground between bolt-action and semi-automatic rifles. They offer quick cycling of rounds while maintaining simplicity. Lever-action rifles are popular for hunting and home defense in rural areas.

Shotguns

Shotguns are highly effective for close-quarters defense and can be used for hunting both small game and larger animals like deer or birds. The ability to use different types of ammunition (such as buckshot, birdshot, or slugs) adds versatility.

Pump-action shotguns: Pump-action shotguns are extremely reliable and easy to maintain. They are excellent for home defense, allowing you to engage multiple targets quickly. The sound of racking a shotgun is also a well-known deterrent.

Break-action shotguns: These are the simplest shotguns, requiring little maintenance and few moving parts. However, their limited capacity (typically two rounds) makes them less effective in prolonged engagements.

Handguns

Handguns are valuable for personal protection, especially when mobility is important. While they lack the range and stopping power of rifles and shotguns, they are easy to carry and conceal.

Semi-automatic pistols: Semi-automatic handguns, such as Glocks or 1911s, are the most common choice for personal defense. They offer a good balance between firepower and portability, with magazine capacities typically ranging from 7 to 17 rounds.

Revolvers: Revolvers are known for their simplicity and reliability. They typically hold fewer rounds than semi-automatic pistols but are easier to maintain and less prone to jamming.

Carbines and Submachine Guns

Carbines, which are shorter versions of rifles, and submachine guns, which fire pistol-caliber rounds, are effective for close-quarters combat and home defense. Their compact size makes them ideal for tight spaces and rapid movement.

Pistol-caliber carbines: These firearms are easy to control, and their smaller caliber rounds (such as 9mm or .45 ACP) are easier to stockpile. They are effective for home defense but lack the range and stopping power of full-sized rifles.

Submachine guns: True submachine guns, such as the Uzi or MP5, are fully automatic or burst-fire weapons, typically reserved for military use. However, civilian-legal semi-automatic versions exist. Their main advantage is portability and high rate of fire, though ammunition consumption can be high.

Ammunition Considerations and Conservation

In a post-apocalyptic world, ammunition may become more valuable than gold. The ability to conserve and manage your ammunition effectively will be critical to long-term survival. Stockpiling the right types of ammunition and learning how to reload spent casings can significantly extend your supply.

Stockpiling Ammunition

When stockpiling ammunition, focus on versatility and availability. Choose common calibers that are widely used, as they will be easier to find or trade for after a disaster.

Common rifle calibers: .223/5.56mm (used in AR-15 rifles), 7.62x39mm (used in AK-47 rifles), and .308 Winchester/7.62 NATO are popular choices. These rounds offer a balance of range, power, and availability.

Common handgun calibers: 9mm, .45 ACP, and .38 Special are widely used and effective for self-defense. Their smaller size also allows you to carry more rounds.

Shotgun shells: 12-gauge and 20-gauge are the most common shotgun calibers. Stockpile a mix of birdshot, buckshot, and slugs for maximum versatility.

Reloading Ammunition

Learning how to reload spent ammunition casings is a valuable skill in a post-apocalyptic world. Reloading allows you to reuse brass casings, provided you have the necessary components: powder, primers, and bullets.

Reloading equipment: Invest in a basic reloading press, powder scales, and dies for your chosen calibers. While the initial investment may be expensive, it can save you money and extend your ammunition supply over time.

Scavenging brass: Even if you don't reload immediately, collect spent brass casings whenever you can. These can be reused or traded with others who know how to reload.

Maintaining and Repairing Firearms

In a world without gunsmiths or regular supply chains, knowing how to maintain and repair your firearms is essential. Firearms that are neglected or poorly maintained will eventually fail, leaving you defenseless.

Cleaning and Maintenance

Regular cleaning is crucial to keep firearms in working order. Dirt, debris, and moisture can cause malfunctions, while neglecting lubrication can lead to rust and wear on metal parts.

Cleaning kits: Invest in a basic cleaning kit that includes brushes, patches, cleaning rods, and oil. These tools will allow you to clean the barrel, action, and moving parts of your firearms.

Lubrication: Keep your firearms properly lubricated, especially in humid or wet environments where rust is a concern. Use gun oil or other lubricants designed for firearms to ensure smooth operation.

Field-stripping: Learn how to field-strip your firearms, which means disassembling them into major components for cleaning and inspection. This skill is particularly important for semi-automatic weapons that can become dirty after extended use.

Simple Repairs and Spare Parts

In a post-apocalyptic world, you may need to repair your firearms without the help of a gunsmith. Keeping a small supply of spare parts for common issues can help you avoid being left with a useless weapon.

Spare parts: For semi-automatic firearms, keep spare parts like firing pins, springs, and extractors, which are the most common components to wear out or break. For bolt-action rifles, spare bolts or bolt parts can be useful.

Improvised repairs: In the absence of spare parts, you may need to improvise repairs. Knowing how to use basic tools and understanding the mechanics of your firearm can help you make temporary fixes in the field.

Ethical and Legal Considerations

While firearms can provide protection and food in a post-apocalyptic world, their use comes with significant ethical and legal considerations. It's important to understand the gravity of using lethal force and to consider the consequences of your actions, both morally and practically.

The Ethics of Self-Defense

In a lawless world, the use of firearms for self-defense may become a necessity, but it should always be a last resort. Taking a life, even in defense, can have profound emotional and psychological effects.

De-escalation first: Whenever possible, attempt to de-escalate confrontations or avoid conflict altogether. Firing a weapon should only be done when there is no other way to protect yourself or others from imminent harm.

Protecting property vs. life: Be cautious about using deadly force to protect property. While it may be tempting to defend your supplies or resources with lethal force, consider the potential consequences and whether it's worth risking lives.

Firearms and Community Dynamics

Firearms can change the dynamic within a survival group or community. While they provide security, they can also create tension if not managed properly.

Training and responsibility: Ensure that anyone in your group who is armed is properly trained in the safe use and handling of firearms. Accidental discharges or improper use can lead to unnecessary injuries or deaths.

Authority and trust: In a post-apocalyptic scenario, communities will need to establish clear rules around who can carry firearms, when they can be used, and how disputes are resolved. Trust is key, and misuse of firearms can quickly erode that trust.

Beyond Firearms: Comprehensive Defense Strategies

While firearms are a valuable tool for defense, they should be part of a larger strategy that includes fortification, vigilance, and cooperation.

Fortifying Your Shelter

Physical fortifications can provide significant defense without the need for constant armed vigilance. Reinforce doors and windows, create barriers, and establish clear sightlines for visibility. The goal is to make it difficult for intruders to enter while giving you time to respond.

Community Defense

In a survival situation, working with others to establish a community defense plan can increase everyone's chances of survival. Pooling resources and skills allows for better protection of people and property.

Shared patrols: Organize patrols or watch shifts to monitor the perimeter of your living area. This reduces the risk of surprise attacks and spreads the burden of security across the group.

Fortified zones: Designate safe areas within your community where vulnerable individuals, such as children and the elderly, can stay protected. Ensure these areas are well-defended and stocked with supplies.

Firearms play a crucial role in post-apocalyptic survival, providing defense against both human and animal threats, as well as enabling hunting for food. However, they are just one part of a comprehensive defense strategy that includes fortification, community cooperation, and responsible use.

Reclaiming the Earth: How Long Until It's Safe Again?

In the aftermath of a nuclear disaster, the question of how long it will take for the Earth to become safe again is a critical one. Radiation from fallout can linger in the environment for years, decades, or even centuries, depending on the type of radioactive materials released. The timeline for recovery varies depending on the severity of the nuclear event, the specific isotopes involved, and the geographical area affected. Understanding how radiation decays, what factors influence the persistence of contamination, and how to navigate the recovery process can help survivors plan for the future.

This chapter will explore the different phases of radiation decay, the environmental impact of fallout, and what can be expected in terms of the timeline for human re-inhabitation and environmental recovery. By understanding the science behind radiation decay and contamination, preppers can better estimate when and where it will be safe to reclaim the Earth after a nuclear disaster.

The Science of Radiation Decay

After a nuclear explosion or accident, radioactive isotopes are released into the atmosphere, soil, and water. These isotopes are unstable and emit radiation as they decay into more stable forms. The rate at which they decay is measured by their half-life, which is the amount of time it takes for half of the radioactive atoms in a substance to disintegrate.

Half-Life and Radioactive Isotopes

Different radioactive isotopes have vastly different half-lives, and understanding which isotopes are present after a nuclear event is key to determining when an area will be safe again.

Iodine-131: One of the more dangerous isotopes released during a nuclear event is iodine-131, which has a half-life of about 8 days. While this isotope decays relatively quickly, it can be extremely harmful in the short term, especially when inhaled or ingested, as it concentrates in the thyroid gland.

Cesium-137: Cesium-137 has a much longer half-life of approximately 30 years. This isotope is highly soluble in water, which means it can spread through the environment and contaminate crops, soil, and water supplies. It is a major concern for long-term contamination of agricultural regions.

Strontium-90: With a half-life of 29 years, strontium-90 behaves similarly to calcium in the human body, accumulating in bones and teeth. This isotope can cause long-term health effects like cancer and bone disorders, making it a persistent environmental threat.

Plutonium-239: One of the most dangerous isotopes, plutonium-239, has a half-life of 24,000 years. While it is not as easily dispersed as cesium or iodine, its extreme toxicity and longevity mean that areas contaminated with plutonium may remain hazardous for millennia.

Phases of Radiation Decay

The process of radiation decay follows a predictable timeline, but the rate of decay and the dangers posed to human life vary depending on the specific isotopes involved.

Immediate fallout (first 48 hours): In the first two days after a nuclear explosion, the radiation levels are at their highest. Fallout particles settle from the atmosphere, contaminating everything they touch. During this phase, exposure to the open air can be fatal, and those who do not seek shelter are at significant risk of acute radiation sickness.

Initial decay (first 2 weeks): Over the next two weeks, radiation levels drop dramatically. For example, radiation levels can decrease by 90% within the first seven hours and by 99% after 48 hours. However, some areas remain highly contaminated, especially those downwind of the blast site.

Short-term recovery (first 3 months): In the months following a nuclear disaster, many of the more volatile radioactive isotopes, such as iodine-131, will have decayed. This reduces the immediate health risks, but longer-lived isotopes like cesium-137 and strontium-90 will still pose significant threats.

Long-term decay (years to centuries): Over the following decades, radiation levels will continue to decrease as cesium-137 and other longer-lived isotopes decay. However, areas heavily contaminated with plutonium or other long-lived isotopes may remain hazardous for centuries or even millennia.

Environmental Impact of Nuclear Fallout

Nuclear fallout affects the environment in several ways, from soil contamination to water pollution. The ability to reclaim the land and live safely in previously contaminated areas depends on the extent of fallout and the type of radioactive material involved.

Soil Contamination

Soil contamination is one of the most long-lasting effects of nuclear fallout. Radioactive isotopes like cesium-137 and strontium-90 bind to soil particles, making it difficult to remove them from the environment. Contaminated soil poses a significant risk to agriculture, as radioactive particles can be absorbed by plants and enter the food chain.

Remediation techniques: Over time, soil can be remediated using techniques such as removing the top layer of contaminated soil, adding clean soil on top, or planting certain types of vegetation that absorb radioactive materials (a process known as phytoremediation). However, these methods are costly and time-consuming, and they are not always effective for long-lived isotopes like plutonium.

Water Contamination

Water contamination is a major concern after a nuclear disaster, as radioactive particles can be carried into rivers, lakes, and groundwater supplies. Cesium-137, in particular, is highly soluble in water and can travel long distances, affecting areas far beyond the initial fallout zone.

Water purification: Purifying contaminated water is possible but requires advanced filtration systems capable of removing radioactive particles. Basic filtration methods, such as boiling, will not remove radiation from water. Distillation and ion-exchange systems are more effective at purifying water, but access to these technologies may be limited in a post-apocalyptic world.

Wildlife and Ecosystem Recovery

Radiation can have devastating effects on local wildlife, causing mutations, reproductive issues, and population declines. However, some ecosystems have shown remarkable resilience in recovering from radiation exposure. For example, wildlife in the Chernobyl Exclusion Zone has rebounded despite the high levels of radiation, as the absence of human activity has allowed animals to repopulate the area.

Mutation and adaptation: While some animals and plants may suffer from radiation-induced mutations, others may adapt to the environment over time. Ecosystems will eventually recover, but this process can take decades or longer, depending on the severity of contamination.

Human Health and Re-Inhabitation

One of the most critical concerns in reclaiming the Earth after a nuclear disaster is when it will be safe for humans to return to contaminated areas. The timeline for safe re-inhabitation depends on radiation levels, the type of radioactive isotopes present, and the effectiveness of decontamination efforts.

Radiation Thresholds for Human Safety

Radiation exposure is measured in units called sieverts (Sv), and health risks increase with the amount of radiation absorbed by the body. Understanding the safe thresholds for radiation exposure is key to determining when an area can be safely re-inhabited.

Short-term safety (1-2 months): Immediately after a nuclear event, radiation levels may be high enough to cause acute radiation sickness with only a short exposure. Once radiation levels drop below 1 sievert per hour, it becomes possible to re-enter an area for short periods of time, but long-term exposure is still dangerous.

Long-term safety (6-12 months): After several months, radiation levels may have dropped enough to allow for limited re-inhabitation, provided that protective measures are taken. An area is generally considered safe for long-term habitation when radiation levels fall below 0.1 sieverts per year. However, living in an area with elevated radiation levels still increases the risk of cancer and other health issues over time.

Health Monitoring and Radiation Protection

Even when an area is deemed safe for re-inhabitation, survivors must remain vigilant about monitoring radiation exposure. Prolonged exposure to low levels of radiation can lead to an increased risk of cancer, thyroid disorders, and other health complications.

Radiation detectors: Personal dosimeters and Geiger counters are essential tools for monitoring radiation levels. These devices allow you to track exposure in real time and avoid highly contaminated areas.

Protective clothing and decontamination: When returning to contaminated areas, wearing protective clothing and regularly decontaminating yourself and your living environment can help reduce exposure. Washing off fallout particles, avoiding contaminated food and water, and staying indoors during periods of high fallout can minimize health risks.

The Timeline for Reclaiming Contaminated Areas

Reclaiming the Earth after a nuclear disaster is a long-term process that requires patience, perseverance, and careful planning. The timeline for safe re-inhabitation and environmental recovery varies depending on the scale of the disaster, the specific isotopes involved, and the effectiveness of remediation efforts.

Short-Term Reclamation (1-3 Years)

In the first few years following a nuclear disaster, the focus will be on stabilizing the environment and decontaminating key areas for human habitation. During this period:

Decontamination efforts: Removing or isolating the most heavily contaminated areas, such as soil removal or water purification, will be critical. This is the period where basic infrastructure may begin to return, and limited re-inhabitation of low-risk areas may occur.

Short-term agriculture: It may be possible to grow food in enclosed or hydroponic systems, where crops are protected from contaminated soil and water. Open-field agriculture will likely be restricted to areas with minimal fallout.

Medium-Term Reclamation (10-30 Years)

Over the next few decades, radiation levels will continue to decrease, and many areas will become safe for human habitation. During this period:

Agriculture and livestock: Farming may resume in areas where radiation levels are low enough to ensure safe food production. However, ongoing testing and monitoring will be necessary to prevent contaminated crops from entering the food supply.

Ecosystem recovery: Wildlife populations will continue to recover, and previously contaminated areas may begin to show signs of natural regeneration. Reforestation and soil remediation efforts can accelerate the recovery process.

Long-Term Reclamation (50-100+ Years)

For the most heavily contaminated areas, such as those affected by plutonium or other long-lived isotopes, full recovery may take centuries. Over the long term:

Abandoned zones: Certain areas may remain off-limits to human habitation for hundreds or even thousands of years due to the presence of long-lived isotopes. These areas may be designated as exclusion zones, much like Chernobyl's Exclusion Zone, where human activity is minimal.

Cultural adaptation: Societies may need to adapt to living in a post-nuclear world, where certain regions are permanently uninhabitable, and others are subject to strict radiation monitoring. Communities may develop new ways of living, such as building elevated or enclosed homes to reduce exposure.

Reclaiming the Earth after a nuclear disaster is a long and complex process, but it is possible with careful planning, decontamination efforts, and an understanding of radiation decay. While some areas may become safe for re-inhabitation within a few years, others will remain hazardous for much longer, particularly those contaminated by long-lived isotopes like plutonium. By monitoring radiation levels, taking protective measures, and focusing on remediation, survivors can gradually reclaim their world and begin the process of rebuilding society in the aftermath of a nuclear event.

The Psychological Toll of Nuclear Winter: Surviving the Mental Strain

A nuclear winter not only threatens survival with extreme environmental conditions, radiation exposure, and scarcity of resources but also poses immense psychological challenges. The mental strain of surviving in a post-apocalyptic world can be as dangerous as the physical threats, as prolonged stress, fear, isolation, and trauma take a heavy toll on survivors. Understanding how to cope with the psychological effects of nuclear winter is critical for maintaining mental health and resilience in the face of overwhelming adversity.

This chapter explores the psychological toll of nuclear winter, focusing on the mental health challenges survivors are likely to face and strategies for managing the mental strain. It covers common emotional reactions, the long-term effects of trauma, and practical techniques to foster resilience and emotional well-being in a devastated world.

Understanding the Psychological Impact of Nuclear Winter

In a world where normal life has been shattered, the psychological impact of nuclear winter is profound. Survivors face a combination of acute trauma, long-term stress, and profound uncertainty about the future. It's essential to recognize these psychological challenges and develop coping mechanisms to deal with them.

Initial Shock and Trauma

The immediate aftermath of a nuclear event can trigger intense psychological shock. Witnessing widespread destruction, losing loved ones, or experiencing life-threatening danger can leave survivors in a state of acute trauma, characterized by feelings of disbelief, numbness, and emotional overwhelm.

Acute stress reaction: In the early stages, many survivors may experience acute stress reactions, which include emotional numbness, difficulty thinking clearly, and physical symptoms such as increased heart rate, sweating, and hypervigilance. These reactions are natural and can be the body's way of protecting itself from immediate psychological harm.

Survivor's guilt: Those who live through a nuclear disaster may experience survivor's guilt, a condition where individuals feel intense guilt or shame for having survived while others did not. This can lead to emotional withdrawal, depression, or self-destructive behaviors.

Long-Term Psychological Stress

As the reality of nuclear winter sets in, the ongoing stress of survival can lead to long-term psychological strain. Factors such as the constant threat of radiation, scarcity of resources, extreme cold, and isolation contribute to a heightened state of anxiety and fear.

Chronic stress: Prolonged exposure to life-threatening conditions and the struggle to secure basic needs like food, water, and shelter can lead to chronic stress. Over time, this can manifest as irritability, fatigue, difficulty concentrating, and physical symptoms like headaches or digestive issues.

Depression and hopelessness: The bleakness of nuclear winter, with its darkened skies, freezing temperatures, and barren landscapes, can foster a sense of hopelessness. Survivors may feel trapped in an unchangeable reality, leading to depression, loss of motivation, and withdrawal from others.

Isolation and Loneliness

In the aftermath of a nuclear disaster, survivors may be cut off from family, friends, and social support networks. Isolation and loneliness can severely impact mental health, as humans are inherently social creatures who rely on connection for emotional well-being.

Social withdrawal: The fear of radiation exposure or dangerous interactions with other survivors may force individuals or families into isolation. While physical isolation may be necessary for survival, prolonged periods of loneliness can exacerbate feelings of despair and hopelessness.

Loss of community: The destruction of social structures, such as neighborhoods, communities, and cities, leaves survivors disconnected from the familiar social fabric of life. The absence of communal rituals, gatherings, and support can intensify feelings of alienation.

Trauma and Post-Traumatic Stress Disorder (PTSD)

Survivors of nuclear winter are likely to experience post-traumatic stress disorder (PTSD), a condition that arises from experiencing or witnessing traumatic events. PTSD can manifest in various ways, including flashbacks, nightmares, emotional numbness, and hyperarousal.

Triggers and flashbacks: Survivors may experience flashbacks or intrusive memories of the initial disaster, triggered by sights, sounds, or situations that remind them of the traumatic event. These flashbacks can cause intense emotional distress and make it difficult to move forward.

Emotional numbness: Some survivors may cope with trauma by shutting down emotionally. This emotional numbness can lead to feelings of detachment, difficulty forming relationships, and a lack of interest in activities that once brought joy.

Survivor Fatigue

As time passes, the ongoing struggle for survival can lead to survivor fatigue—a state of physical, emotional, and mental exhaustion. Constantly living in "survival mode," where every day is a battle to stay alive, wears down resilience and leads to burnout.

Mental exhaustion: The need to make life-or-death decisions daily, combined with the continuous stress of resource scarcity and environmental hazards, can lead to cognitive overload and mental fatigue. Survivors may find it difficult to focus, make decisions, or even maintain hope.

Emotional numbness or outbursts: As survivor fatigue sets in, some may experience emotional detachment, while others may have sudden emotional outbursts due to the buildup of stress and frustration.

Coping Strategies for Managing the Mental Strain

While the psychological toll of nuclear winter is severe, there are strategies survivors can employ to manage mental strain and foster emotional resilience. Developing coping mechanisms, building community, and prioritizing mental health can help individuals navigate the challenges of life in a post-nuclear world.

Establishing Routine and Structure

One of the most effective ways to combat the chaos and uncertainty of nuclear winter is to establish a daily routine. Creating structure in your day provides a sense of normalcy and control, which can help alleviate feelings of helplessness and anxiety.

Daily tasks: Even in survival situations, setting aside time for basic tasks like cleaning, organizing supplies, or preparing meals can provide a sense of accomplishment and purpose. It also helps break the monotony of endless survival.

Setting goals: Establishing small, achievable goals each day—such as gathering firewood, fixing shelter, or rationing food—gives survivors a sense of progress. Focusing on immediate, tangible objectives helps redirect attention from the overwhelming reality of nuclear winter.

Maintaining Social Connections

Human connection is essential for mental well-being, even in the direst of circumstances. Prioritizing social interaction, whether within a small group of survivors or through more creative means, can mitigate feelings of loneliness and isolation.

Group support: If you are surviving with others, make time for conversation, teamwork, and mutual support. Sharing the burden of survival and working together can create a sense of solidarity and reduce emotional isolation.

Creative communication: In the absence of a physical community, finding ways to communicate with others—through radios, handwritten notes, or other forms of outreach—can foster a sense of connection. Even brief communication with others can provide emotional relief.

Practicing Mindfulness and Meditation

Mindfulness and meditation are powerful tools for managing stress, anxiety, and trauma. While it may seem counterintuitive in a survival situation, taking time to center yourself mentally can help you remain focused, calm, and clear-headed in the face of danger.

Breathing exercises: Simple breathing exercises, such as slow, deep breathing, can help reduce anxiety and lower stress levels. Taking just a few minutes each day to focus on your breath can bring a sense of calm, even in the midst of chaos.

Grounding techniques: Grounding techniques, which involve focusing on physical sensations or your surroundings, can help survivors manage overwhelming emotions. For example, focusing on the texture of an object, the sound of the wind, or the feel of the ground underfoot can bring you back to the present moment and reduce panic.

Creative Outlets and Self-Expression

Finding ways to express emotions creatively can be a therapeutic outlet for processing trauma, grief, and fear. Even in a survival scenario, taking time for self-expression can help alleviate emotional burdens.

Journaling: Writing about your experiences, thoughts, and feelings can help you process the psychological strain of survival. Journaling allows you to express emotions that might be difficult to talk about and provides a sense of release.

Art and music: Drawing, painting, or playing music—even with limited resources—can provide an emotional escape and a way to process complex feelings. These creative activities offer a reprieve from the harshness of daily survival and can be a shared activity with others.

Building Resilience and Mental Toughness

Resilience is the ability to adapt to adversity and recover from difficult situations. In a post-nuclear world, developing resilience is essential for long-term survival, both mentally and physically.

Focus on what you can control: A key component of resilience is focusing on what is within your control, rather than dwelling on the uncertainty or uncontrollability of the larger situation. Identify the areas where you can take action—whether it's gathering supplies, securing shelter, or taking care of your health—and direct your energy there.

Finding purpose: Even in the darkest times, finding a sense of purpose can sustain mental resilience. Whether it's protecting loved ones, rebuilding a community, or simply surviving another day, having a goal can keep you motivated and engaged in life.

Dealing with Grief and Loss

Survivors of a nuclear disaster are likely to experience significant grief and loss, whether from the death of loved ones, the destruction of their homes, or the loss of their way of life. Processing grief in a healthy way is crucial for mental health.

Acknowledging grief: It's important to allow yourself to feel grief rather than suppress it. Acknowledging the loss and giving yourself permission to mourn can prevent emotional numbness or future breakdowns.

Honoring memories: Finding ways to honor those you've lost—whether through rituals, storytelling, or creating a physical memorial—can provide closure and help you process the grief.

Seeking Professional Help (If Available)

While professional mental health support may be difficult to find in a post-apocalyptic world, if you have access to a therapist, counsellor, or psychologist, seeking their help can be invaluable. They can provide coping strategies, trauma therapy, and emotional support to navigate the psychological challenges of nuclear winter.

Peer support: In the absence of professional mental health services, peer support can fill the gap. Sharing experiences and emotional struggles with other survivors creates a network of mutual understanding and solidarity.

Recognizing Signs of Mental Distress

Being aware of the signs of mental distress in yourself and others is crucial for preventing serious psychological issues from worsening. Early intervention can make a significant difference in mental health outcomes.

Signs of Depression

Symptoms of depression can include persistent sadness, loss of interest in activities, fatigue, changes in appetite or sleep patterns, and thoughts of hopelessness or suicide. In a survival situation, these symptoms can severely hinder your ability to function, making it important to address them early.

Signs of Anxiety

Anxiety often manifests as restlessness, irritability, difficulty concentrating, muscle tension, or panic attacks. Chronic anxiety can impair decision-making and lead to poor judgment, which is dangerous in a survival setting.

Signs of PTSD

PTSD symptoms include flashbacks, nightmares, hypervigilance, emotional numbness, and avoidance of situations that remind the individual of the trauma. If you or someone else is exhibiting signs of PTSD, it's important to practice grounding techniques and seek help if available.

The psychological toll of nuclear winter is immense, but understanding the mental health challenges and developing effective coping strategies can help you survive not just physically but emotionally. The mental strain of isolation, trauma, and prolonged stress is unavoidable in a post-apocalyptic world, but by building resilience, maintaining social connections, and taking care of your mental health, you can navigate the harsh realities of nuclear winter with greater strength and endurance. Above all, fostering hope, purpose, and connection is key to surviving the psychological burdens of life after a nuclear disaster.

Final Thoughts on Surviving the Nuclear Winter

Surviving a nuclear winter requires more than just the physical endurance to withstand extreme conditions; it demands mental fortitude, careful planning, and a deep understanding of the challenges ahead. The combination of radioactive fallout, extreme cold, scarcity of resources, and social collapse presents one of the most daunting survival scenarios imaginable. Yet, survival is possible with the right mindset, knowledge, and preparation.

As we conclude our exploration of surviving a nuclear winter, it's important to reflect on the key takeaways that can guide you through the most difficult aspects of such a scenario. Whether you are preparing for the worst or just seeking to understand what survival would require, these final thoughts can serve as a blueprint for how to navigate the devastating aftermath of a nuclear event.

Prepare to Face the Immediate Threats

The first step in surviving a nuclear winter is to address the immediate dangers: radiation, fallout, and the initial chaos following a nuclear event. Understanding how to protect yourself from radiation exposure and ensuring you have adequate shelter, food, and water supplies is essential in the first few days and weeks.

Find or create adequate shelter: Your first priority should be finding a space that provides protection from radiation and extreme cold. Whether it's an underground bunker, a fortified basement, or a well-shielded safe room, the quality of your shelter will determine your chances of surviving the initial fallout.

Stockpile essential resources: Food, water, and medical supplies are the foundation of long-term survival. Focus on acquiring non-perishable items, radiation protection tools (such as potassium iodide tablets and Geiger counters), and materials to purify water and decontaminate surfaces.

Avoid immediate fallout exposure: The first 48 hours after a nuclear event are the most dangerous. Remain sheltered during this critical period and take steps to decontaminate yourself if exposed to radioactive particles.

Develop a Long-Term Survival Strategy

While immediate survival is critical, nuclear winter is a long-term challenge. The extreme environmental changes caused by nuclear winter, including plummeting temperatures, prolonged darkness, and disruptions to agriculture, require a sustainable approach to survival.

Plan for food production: Traditional farming may be impossible, but hydroponics, indoor gardening, and raising small livestock can provide a source of food in a fallout world. You'll need to adapt your food production techniques to avoid contamination and maximize limited space and resources.

Create a sustainable water supply: Water sources may become contaminated, so developing a reliable filtration system or collecting rainwater is vital. Ensure you have the tools and knowledge to purify water continuously over the long term.

Conserve energy and resources: In the harsh conditions of nuclear winter, conserving energy will be critical to survival. Whether it's rationing food and water or finding alternative ways to heat your shelter, careful management of limited resources is essential for long-term endurance.

Stay Informed and Adapt

Survival in a nuclear winter is about adaptation. The world you once knew may no longer exist, and flexibility in your plans will determine your success. Staying informed about your environment, radiation levels, and changing conditions is vital to making smart decisions.

Use tools to monitor radiation: Radiation levels will fluctuate, and it's crucial to know when it's safe to go outside, forage, or hunt. A reliable radiation detector will be your lifeline for making decisions about movement and resource gathering.

Adapt to changing conditions: The extreme cold and lack of sunlight during nuclear winter can make everyday tasks more challenging. Be prepared to adapt to low-light environments, freezing temperatures, and shifting landscapes. Your ability to stay flexible and resourceful will be key.

Mental Resilience is Key

Perhaps the most difficult challenge in surviving a nuclear winter is maintaining your mental health and emotional resilience. Isolation, loss, trauma, and the bleakness of the environment can take a heavy psychological toll. Developing mental strength is as important as physical preparation.

Stay connected if possible: Human connection, whether with family, a survival group, or even through radio communication, can provide emotional support and reduce feelings of isolation. Share the burden of survival and work together to create a sense of community and purpose.

Practice emotional self-care: Make time to process grief, fear, and stress in a healthy way. Simple routines, creative outlets, and mindfulness practices can help you manage overwhelming emotions and maintain hope.

Focus on small victories: Survival in such an extreme situation can feel like a never-ending struggle, but focusing on small successes—securing clean water, gathering food, fortifying shelter—can boost morale and provide motivation to keep going.

The Role of Community and Collaboration

In the long term, surviving and rebuilding in the aftermath of a nuclear winter may require collaboration with others. The isolation and scarcity of resources may drive people together, either to form new communities or rebuild existing ones. While there may be risks involved in interacting with others, especially in a chaotic, lawless environment, cooperation can increase your chances of survival.

Establish mutual aid networks: Sharing skills, knowledge, and resources within a group can make survival tasks more manageable. A group offers security, division of labor, and the opportunity to rebuild a sense of normalcy.

Exercise caution with strangers: Trust is essential in a post-apocalyptic world, but it must be earned. Carefully assess those outside your immediate circle before deciding to collaborate or share resources.

Plan for long-term recovery: While survival is the immediate focus, it's also important to consider how to rebuild society in the future. Creating plans for restoring agriculture, education, and infrastructure can lay the groundwork for a new beginning.

Fostering Hope for the Future

Nuclear winter paints a bleak picture of the future, but even in the most dire circumstances, hope and the human spirit's resilience can shine through. Rebuilding a world after a nuclear disaster may seem impossible, but focusing on small, tangible steps toward recovery can provide a sense of purpose.

Look to history for lessons: Past nuclear disasters like Chernobyl and Fukushima have shown that nature and humanity can recover over time, even in the face of extreme radiation. While nuclear winter is a much larger and more destructive event, these historical examples offer hope for long-term environmental recovery.

Focus on the next generation: If children or future generations are part of your group, teaching them survival skills, preserving knowledge, and instilling values of resilience and cooperation will be critical for rebuilding society after the immediate crisis has passed.

Surviving a nuclear winter is an extraordinary challenge, but it is not impossible. By preparing both mentally and physically, adapting to changing conditions, and focusing on long-term sustainability, you can increase your chances of enduring this catastrophic event. While the road to recovery will be long and fraught with difficulties, fostering hope, community, and resilience will be key to navigating the harsh realities of a post-nuclear world.

The human spirit, tested by adversity throughout history, has shown time and time again that survival is possible in even the most extreme circumstances. The journey through nuclear winter will test every aspect of your being—your resourcefulness, your strength, and your will to live—but with preparation and perseverance, you can emerge on the other side, ready to reclaim a new world.

A Comprehensive Two Year Nuclear Winter Survival List of Food and Equipment

Surviving a nuclear winter requires careful preparation and planning, particularly when it comes to the essential food and equipment needed to sustain life for an extended period of time. In the aftermath of a nuclear event, resources will become scarce, and traditional means of obtaining food, water, and supplies will be limited or impossible. This chapter outlines a comprehensive survival list of food and equipment that will help you prepare for the harsh conditions of nuclear winter, ensuring that you have everything you need to survive for two years.

The Importance of Long-Term Planning

A nuclear winter, characterized by extreme cold, reduced sunlight, and widespread contamination, presents unique survival challenges. Preparing for two years of survival means that you must focus on long-term sustainability, rationing, and careful resource management. Your stockpile should not only cover your basic needs but also take into account the difficulty of replenishing supplies and the health impacts of prolonged survival in such an environment.

The following is a detailed list of food and equipment that will provide the foundation for your long-term survival. The list provided below is primarily designed for one person for a two-year survival period. However, it can easily be scaled to accommodate additional people by multiplying the quantities based on the number of individuals in your group.

For example, if you're preparing for a family of four, you would roughly multiply most food and essential quantities by four, although some items—like cooking equipment or tools—might not need to be duplicated and can be shared among the group.

It's also important to remember that individual needs vary, especially in terms of calorie intake (depending on age, gender, and activity level), so you may need to adjust the quantities slightly based on your group's specific needs. Additionally, it's always a good idea to have a little extra in case of unexpected emergencies or delays in restocking supplies.

Essential Food Supplies for 2 Years

The food you stockpile must be non-perishable, calorie-dense, and capable of providing essential nutrients over an extended period. You will need a balance of carbohydrates, proteins, fats, vitamins, and minerals to maintain health in harsh conditions.

Carbohydrates (Staples for Energy)

Carbohydrates are the primary source of energy, and in a survival situation, you'll need a variety of grains, starches, and calorie-rich foods.

Rice (300 lbs): Long-grain white rice has an almost indefinite shelf life and provides a solid base for many meals.

Pasta (150 lbs): Dried pasta is easy to store and prepare. It can be mixed with various ingredients to create calorie-dense meals.

Oats (100 lbs): Oats are a versatile grain that can be used for breakfast, baking, or even as a filler in other dishes.

Dried Beans and Lentils (200 lbs): Beans and lentils are packed with protein and fiber. They're essential for both nutrition and long-term storage.

Flour (100 lbs): Stored in airtight containers, flour can be used to bake bread, make noodles, or create other staples. Wheat flour will last up to a year, but flour alternatives such as almond or coconut flour may have a longer shelf life.

Proteins (Building Blocks of Muscle and Vitality)

Protein is essential for maintaining muscle mass and overall health, particularly in physically demanding survival situations.

Canned Meat (200 cans): Stockpile canned beef, chicken, tuna, and salmon. These protein sources have long shelf lives and are easy to prepare.

Dried Meat (100 lbs): Jerky or dried fish can provide a convenient source of protein that doesn't require refrigeration.

Powdered Eggs (50 lbs): Powdered eggs are lightweight, have a long shelf life, and provide a source of protein and fat.

Nuts and Seeds (50 lbs): Almonds, sunflower seeds, and peanuts are high in protein and healthy fats. These should be vacuum-sealed to prevent spoilage.

Protein Powder (20 lbs): A backup option for quick protein intake, especially if fresh or canned meat supplies are running low.

Fats (Energy-Dense and Essential for Survival)

Fats are crucial for long-term survival, providing concentrated energy and helping the body absorb vitamins.

Vegetable Oil (20 gallons): Use for cooking and adding calories to meals. Olive oil and coconut oil also have longer shelf lives.

Peanut Butter (50 lbs): High in calories, protein, and fat, peanut butter is a survival staple.

Lard or Ghee (10 lbs): These shelf-stable fats can be used for frying, baking, or adding flavor to meals.

Fruits and Vegetables (Vitamins and Fiber)

It's crucial to have a supply of fruits and vegetables to avoid nutrient deficiencies.

Canned Vegetables (300 cans): Stock up on a variety of canned vegetables like peas, carrots, green beans, and tomatoes.

Canned Fruits (200 cans): Fruit is a source of vitamins and natural sugars. Focus on peaches, pineapples, pears, and applesauce.

Dried Fruits (50 lbs): Dried fruits like raisins, apricots, and apples provide a long-lasting source of vitamins and sugars.

Dehydrated Vegetables (50 lbs): These can be rehydrated and used in soups, stews, and casseroles.

Dairy and Alternatives

While fresh dairy won't be available, there are several long-lasting alternatives.

Powdered Milk (100 lbs): An essential source of calcium, protein, and fat, powdered milk can be used in cooking or as a beverage.

Cheese Powder (20 lbs): Cheese powder can add flavor and fat to meals and has a long shelf life.

Shelf-Stable Almond or Soy Milk (40 cartons): These are useful for cooking or as a milk replacement.

Baking and Cooking Essentials

These ingredients will allow you to cook a variety of meals and maintain a semblance of normalcy in your diet.

Baking Powder and Baking Soda (10 lbs each): Essential for making bread, pancakes, and other baked goods.

Sugar (100 lbs): Adds calories to your diet and is essential for baking and preserving food.

Salt (50 lbs): Used for preserving food and seasoning.

Spices (various): Stock up on a variety of spices to enhance flavor and maintain morale.

Essential Equipment for 2 Years of Survival

Food alone won't ensure survival; you'll need a range of equipment to manage food preparation, water purification, shelter, and personal safety. Below is a list of critical survival gear.

Cooking Equipment

Propane Stove and Fuel (50 small propane tanks): Reliable, portable, and essential for cooking when electricity is unavailable.

Cast Iron Cookware: Durable and versatile, cast iron can be used on an open fire or stovetop.

Manual Can Opener: Ensure you have several sturdy can openers for accessing your canned goods.

Solar Oven: A solar oven allows you to cook food using only sunlight, making it a valuable resource when fuel is scarce.

Water Purification and Storage

Water contamination is a major concern after a nuclear disaster. These items will help you secure clean drinking water.

Water Filters (2 high-capacity filters): Gravity-based water filtration systems (such as a Berkey filter) can provide safe drinking water.

Water Purification Tablets (500 tablets): Use these in conjunction with filters to purify water from questionable sources.

Collapsible Water Containers (10 gallons total capacity): For storing water, these containers are lightweight and space-efficient.

Rainwater Collection System: Set up a system to collect rainwater for drinking, washing, and irrigation.

Heating and Shelter

With temperatures dropping during nuclear winter, maintaining warmth is a top priority.

Wood Stove or Rocket Stove: A wood stove will be essential for heating your shelter and cooking. Rocket stoves are portable, efficient, and require minimal fuel.

Thermal Blankets (10): Space blankets help retain body heat and are lightweight.

Winter Clothing: Stockpile thermal underwear, wool socks, insulated boots, and waterproof outerwear for every member of your household.

Insulating Materials: Use tarps, plastic sheeting, and insulation materials to fortify your shelter.

Medical Supplies

Maintaining health over two years requires more than just first aid; you'll need a range of medical supplies.

First Aid Kit: A comprehensive kit with bandages, antiseptics, painkillers, and wound care items.

Prescription Medications: If possible, stockpile essential medications for chronic conditions.

Over-the-Counter Medications: Include pain relievers, antihistamines, cold medicine, and anti-diarrheal pills.

Radiation Sickness Treatments: Stock up on potassium iodide tablets and other treatments for radiation exposure.

Tools and Miscellaneous Equipment

These tools will be critical for survival tasks, repairs, and maintaining your shelter.

Multitool: A high-quality multitool can help with a variety of tasks, from repairs to food preparation.

Axe and Saw: Essential for cutting firewood, building, and general repairs.

Fire Starters (100): Stockpile matches, lighters, and fire-starting kits.

Hand-Crank Radio and Flashlights: Stay informed and keep your space lit without relying on batteries.

Solar Charger: For charging small devices like radios, flashlights, and batteries.

Duct Tape and Rope: Useful for repairs, securing tarps, and building structures.

Conclusion: Enduring and Thriving in the Aftermath

Surviving a nuclear winter is an unprecedented challenge, requiring a combination of physical preparation, mental resilience, and community cooperation. As we've explored throughout this book, survival in such extreme conditions is about more than just stockpiling supplies or building a shelter. It's about adapting to a changed world, maintaining hope in the face of overwhelming adversity, and planning for both immediate survival and long-term sustainability.

Key Principles of Survival

If there's one key takeaway from this book, it's that survival is a multifaceted challenge. Your ability to endure a nuclear winter will depend on how well you prepare, how quickly you can adapt, and how you cope with the psychological and physical demands of a post-apocalyptic environment. By understanding the various elements we've covered—radiation protection, food and water security, mental health, community building, and long-term recovery—you'll be better equipped to face the reality of life after a nuclear event.

Immediate Action Saves Lives

The first hours and days following a nuclear disaster are the most critical. Knowing how to protect yourself from immediate fallout, radiation exposure, and the chaos that follows is essential. Seeking proper shelter, understanding how to decontaminate yourself and your surroundings, and staying informed through whatever communication channels are available will give you the best chance of surviving those early, dangerous days.

Key Takeaway: Having a pre-established plan and the right tools—such as a Geiger counter, potassium iodide tablets, and an emergency shelter—can mean the difference between life and death in the immediate aftermath.

Self-Sufficiency is the Goal

As the nuclear winter drags on, your focus will shift from immediate survival to long-term sustainability. Stockpiling enough food, water, and medical supplies for two years is essential, but the ultimate goal is to achieve a level of self-sufficiency that will carry you through even longer if necessary. This means learning how to grow food indoors, purify water, and make the most of limited resources.

Key Takeaway: The more self-sufficient you can become—whether through hydroponic gardening, raising small livestock, or harnessing alternative energy sources—the more independent and secure your survival will be.

Mental Resilience is Critical

While physical preparation is vital, surviving the psychological strain of nuclear winter may be even more challenging. Isolation, trauma, and the relentless stress of daily survival can take a heavy toll on your mental health. Maintaining social connections, even with a small group, practicing mindfulness, and finding ways to cope with the emotional burden are crucial for your long-term mental well-being.

Key Takeaway: Mental toughness is just as important as physical preparedness. Regularly checking in on your emotional state, fostering hope, and focusing on small victories can help you maintain the will to survive.

Community and Cooperation Increase Chances of Survival

While there are risks associated with interacting with others in a post-apocalyptic world, community and cooperation often provide the greatest security in the long term. Sharing skills, resources, and labor within a trusted group of survivors can dramatically increase your chances of survival. It allows you to pool knowledge and supplies, as well as establish a support network that can help you cope with the physical and emotional demands of a nuclear winter.

Key Takeaway: Build a reliable group of survivors or tap into existing networks. A strong community will give you access to more resources, provide emotional support, and increase your collective resilience in the face of adversity.

Planning for Recovery and Rebuilding

Survival isn't just about enduring the present—it's about planning for the future. Even in the face of nuclear winter, there will come a time when the immediate dangers recede, and thoughts of rebuilding emerge. Your ability to plan for recovery—by preserving knowledge, safeguarding tools, and keeping the spark of hope alive—will determine how well you can transition from survival to rebuilding.

Key Takeaway: Rebuilding society after a nuclear event will take time, effort, and ingenuity. Whether you're preserving seeds for future crops, maintaining equipment for future use, or creating strategies for community rebuilding, the steps you take now will lay the foundation for a future beyond nuclear winter.

The Path Forward

Life After Midnight is not just a survival guide—it's a testament to human resilience. History has shown that even in the darkest times, humanity can endure, adapt, and rebuild. While nuclear winter presents extreme and unprecedented challenges, the principles outlined in this book provide a roadmap for surviving those challenges and, eventually, thriving again.

The path forward will not be easy, but with the right mindset, preparation, and skills, you can face the unknown with confidence. Whether you are a seasoned prepper or someone just beginning to think about survival, remember that knowledge, planning, and adaptability are your most powerful tools.

The world may change irrevocably, but as long as you hold on to the will to survive, the knowledge you've gained here will guide you through the trials ahead. The night may be long and cold, but dawn will come again.

Final Thoughts

Nuclear winter is a worst-case scenario that none of us ever want to face, but if we do, being prepared is the best way to protect yourself, your loved ones, and your future. Take the steps outlined in this book seriously, build your stockpiles, sharpen your skills, and cultivate the resilience needed to face whatever comes next.

In the end, survival is not just about staying alive—it's about creating the possibility of a future. And no matter how dark it seems after midnight, with preparation and determination, a new day will eventually come.

Don't miss out!

Visit the website below and you can sign up to receive emails whenever Andrew Parry publishes a new book. There's no charge and no obligation.

https://books2read.com/r/B-A-FROLC-DLNBF

BOOKS 2 READ

Connecting independent readers to independent writers.

About the Author

Andrew Parry is a dedicated researcher and author specializing in non-fiction works. His writing is driven by a deep passion for exploring the subjects that resonate with his core beliefs and concerns. Andrew's work reflects his profound commitment to addressing some of the most pressing issues facing humanity today, including the future of our species, the environment, and the looming threat of extinction.